Learn RStudio IDE

Quick, Effective, and Productive Data Science

Matthew Campbell

Apress®

Learn RStudio IDE: Quick, Effective, and Productive Data Science

Matthew Campbell
Yardley, PA, USA

ISBN-13 (pbk): 978-1-4842-4510-1   ISBN-13 (electronic): 978-1-4842-4511-8
https://doi.org/10.1007/978-1-4842-4511-8

Copyright © 2019 by Matthew Campbell

Managing Director, Apress Media LLC: Welmoed Spahr
Acquisitions Editor: Steve Anglin
Development Editor: Matthew Moodie
Coordinating Editor: Mark Powers

Cover designed by eStudioCalamar

Cover image designed by Freepik (www.freepik.com)

Distributed to the book trade worldwide by Springer Science+Business Media New York, 233 Spring Street, 6th Floor, New York, NY 10013. Phone 1-800-SPRINGER, fax (201) 348-4505, e-mail orders-ny@springer-sbm.com, or visit www.springeronline.com. Apress Media, LLC is a California LLC and the sole member (owner) is Springer Science + Business Media Finance Inc (SSBM Finance Inc). SSBM Finance Inc is a **Delaware** corporation.

For information on translations, please e-mail editorial@apress.com; for reprint, paperback, or audio rights, please email bookpermissions@springernature.com.

Apress titles may be purchased in bulk for academic, corporate, or promotional use. eBook versions and licenses are also available for most titles. For more information, reference our Print and eBook Bulk Sales web page at http://www.apress.com/bulk-sales.

Any source code or other supplementary material referenced by the author in this book is available to readers on GitHub via the book's product page, located at www.apress.com/9781484245101. For more detailed information, please visit http://www.apress.com/source-code.

Printed on acid-free paper

Table of Contents

About the Author

Matthew Campbell is a research data analyst who has worked on data problems in educational psychology, psychometrics and other research areas for over 15 years. Matt is passionate about technology which he uses to create stories with data, find insights that inform analysis and solves problems for businesses. He received his Masters in Management of Information Systems from the University of Phoenix and Bachelors in Psychology from Pennsylvania State University. Matt has authored 4 books on mobile app development, lead coding bootcamps and worked on various large-scale research projects.

About the Technical Reviewer

 **Dr. Prachee Chaturvedi** is Data Scientist and Digital Strategy Lead in Regulatory, R&D, in Bayer and was a member of Monsanto's Leadership Development Program. She received her PhD in Biological Engineering from University of Florida, Gainesville, FL, and Bachelors in Mechanical Engineering from HBTI, Kanpur, India. Currently, Prachee is focused on establishing comprehensive digital strategy for Regulatory Sciences and developing digital tools for automated and on-demand analysis for Regulatory Safety Data Submission at Bayer Crop Science R&D. She has authored 15 scientific publications, 2 book chapters, and led various cross functional scientific projects.

CHAPTER 1

Installing RStudio

RStudio is an integrated development environment (IDE) that adds modern features like syntax highlighting and code refactoring to R. The strength of RStudio is that it brings all the features that you need together in one place. Before we install RStudio, we will need to install the latest version of R for our operating system as well as another program called git. Both of these software packages are integrated into RStudio, but they do not come with the RStudio IDE.

R is a statistical programming language and we need this language to use RStudio. Git is a popular version control system that will be integrated into RStudio. Version control systems like git are used to manage copies of the code you are working on. Git helps you see the differences between versions of a file so that you can get insight into why one version of your code behaves differently than another version. Git is also required to use Github, an online community where programmers share code and projects.

Install R

R is a free open source software used for statistical programming and graphics. You can get the latest version of R from CRAN, the Comprehensive R Network. This is a network of mirrored servers that have copies of R as well as R packages. R packages are extensions to the core R programming language contributed by R users in the community.

© Matthew Campbell 2019
M. Campbell, *Learn RStudio IDE*, https://doi.org/10.1007/978-1-4842-4511-8_1

To install R, go to this URL: `https://cran.r-project.org/mirrors.html` and you will see a list like the one featured in Figure 1-1.

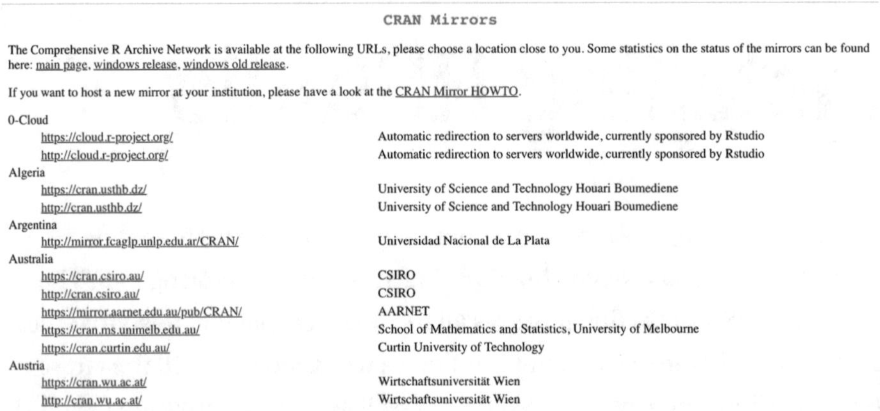

CRAN Mirrors

The Comprehensive R Archive Network is available at the following URLs, please choose a location close to you. Some statistics on the status of the mirrors can be found here: main page, windows release, windows old release.

If you want to host a new mirror at your institution, please have a look at the CRAN Mirror HOWTO.

0-Cloud
    https://cloud.r-project.org/   Automatic redirection to servers worldwide, currently sponsored by Rstudio
    http://cloud.r-project.org/   Automatic redirection to servers worldwide, currently sponsored by Rstudio
Algeria
    https://cran.usthb.dz/   University of Science and Technology Houari Boumediene
    http://cran.usthb.dz/   University of Science and Technology Houari Boumediene
Argentina
    http://mirror.fcaglp.unlp.edu.ar/CRAN/   Universidad Nacional de La Plata
Australia
    https://cran.csiro.au/   CSIRO
    http://cran.csiro.au/   CSIRO
    https://mirror.aarnet.edu.au/pub/CRAN/   AARNET
    https://cran.ms.unimelb.edu.au/   School of Mathematics and Statistics, University of Melbourne
    https://cran.curtin.edu.au/   Curtin University of Technology
Austria
    https://cran.wu.ac.at/   Wirtschaftsuniversität Wien
    http://cran.wu.ac.at/   Wirtschaftsuniversität Wien

Figure 1-1. *CRAN mirrors*

Scroll down the list of servers and find the one that is geographically closest to you and then click the link. Don't worry too much about getting the absolute closest server, just picking any from your country should be fine. If you find a broken link, simply go back and try another server. You will be brought to a new screen that lists versions of R for Windows and Mac as you can see in Figure 1-2.

The Comprehensive R Archive Network

Download and Install R

Precompiled binary distributions of the base system and contributed packages, **Windows and Mac** users most likely want one of these versions of R:

- Download R for Linux
- Download R for (Mac) OS X
- Download R for Windows

R is part of many Linux distributions, you should check with your Linux package management system in addition to the link above.

Source Code for all Platforms

Windows and Mac users most likely want to download the precompiled binaries listed in the upper box, not the source code. The sources have to be compiled before you can use them. If you do not know what this means, you probably do not want to do it!

- The latest release (2018-07-02, Feather Spray) R-3.5.1.tar.gz, read what's new in the latest version.

- Sources of R alpha and beta releases (daily snapshots, created only in time periods before a planned release).

- Daily snapshots of current patched and development versions are available here. Please read about new features and bug fixes before filing corresponding feature requests or bug reports.

- Source code of older versions of R is available here.

- Contributed extension packages

CRAN
Mirrors
What's new?
Task Views
Search

About R
R Homepage
The R Journal

Software
R Sources
R Binaries
Packages
Other

Documentation
Manuals
FAQs
Contributed

Figure 1-2. *Installing R on CRAN*

Click the link for your operating system. Each operating system has different instructions to set up R.

Installing R on Mac

If you are installing R on Mac, you will be presented with a lot of detail about the different R releases. Most people will simply go to the area toward the middle of the page under the heading Latest release as shown in Figure 1-3: and then click the file that ends in with the .pkg file extension. The exact name of the file will be different based on the most current release of the installer.

R for Mac OS X

This directory contains binaries for a base distribution and packages to run on Mac OS X (release 10.6 and above). Mac OS 8.6 to 9.2 (and Mac OS X 10.1) are no longer supported but you can find the last supported release of R for these systems (which is R 1.7.1) here. Releases for old Mac OS X systems (through Mac OS X 10.5) and PowerPC Macs can be found in the old directory.

Note: CRAN does not have Mac OS X systems and cannot check these binaries for viruses. Although we take precautions when assembling binaries, please use the normal precautions with downloaded executables.

As of 2016/03/01 package binaries for R versions older than 2.12.0 are only available from the CRAN archive so users of such versions should adjust the CRAN mirror setting accordingly.

CRAN
Mirrors
What's new?
Task Views
Search

About R
R Homepage
The R Journal

Software
R Sources
R Binaries
Packages
Other

Documentation
Manuals
FAQs
Contributed

R 3.5.1 "Feather Spray" released on 2018/07/05

Important: since R 3.4.0 release we are now providing binaries for OS X 10.11 (El Capitan) and higher using non-Apple toolkit to provide support for OpenMP and C++17 standard features. To compile packages you may have to download tools from the tools directory and read the corresponding note below.

Please check the MD5 checksum of the downloaded image to ensure that it has not been tampered with or corrupted during the mirroring process. For example type
md5 R-3.5.1.pkg
in the *Terminal* application to print the MD5 checksum for the R-3.5.1.pkg image. On Mac OS X 10.7 and later you can also validate the signature using
pkgutil --check-signature R-3.5.1.pkg

Lastest release:

R-3.5.1.pkg
MD5-hash: 58eaf0b5bd024f267ef1e521e17e7f8
SHA1-
hash: 76c01bfa62a6896d5f4a4511e25d17276d149621
(ca. 74MB)

R 3.5.1 binary for OS X 10.11 (El Capitan) and higher, signed package. Contains R 3.5.1 framework, R.app GUI 1.70 in 64-bit for Intel Macs, Tcl/Tk 8.6.6 X11 libraries and Texinfo 5.2. The latter two components are optional and can be ommitted when choosing "custom install", they are only needed if you want to use the tcltk R package or build package documentation from sources.

Figure 1-3. *Mac R installer*

Click the pkg link to install the most recent version of R. Note that if you have a special need for an older version of R, you can scroll down to see options on installing previous version of R. The pkg file will appear in your Downloads folder. Go to your Downloads folder and click the pkg file to install R. A screen like Figure 1-4 will appear.

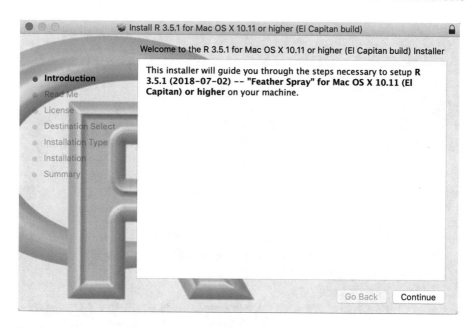

Figure 1-4. *Installing R on Mac*

You will get a series of screens like the one in Figure 1-4. Click Continue for each screen. At one point, a model dialog box will pop up asking for you to agree to the software license. Click Agree to move on. Next, you will be presented with an option to install to a specific disk as shown in Figure 1-5.

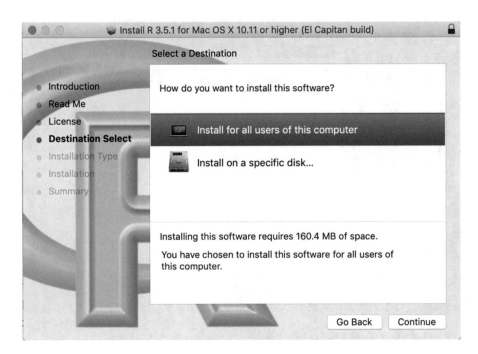

Figure 1-5. *Choosing R location*

Most people will simply choose the option to install for all users.
Choose the first option and click the Continue button. At the next screen,
click the Install button. You will be prompted for your Mac password since
the R installer will make changes to your Mac. Type that in to continue. R
will install, and when everything is done you should see a screen like in
Figure 1-6.

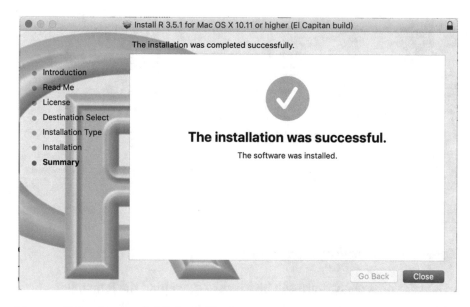

Figure 1-6. *Successful R installation*

Click the Close button and R will be installed on your Mac. You will find R in your Applications folder. Double click the R.app program in your Applications folder to verify that you have R installed and working. The native R Console screen will appear. You can play around with the system here if you already know a little R, or you can simply exit the program. We will revisit using R once RStudio is installed.

Installing R on Windows

Once you click the R for Windows link, you will see a page of instructions. Click on "install R for the first time" as shown in Figure 1-7.

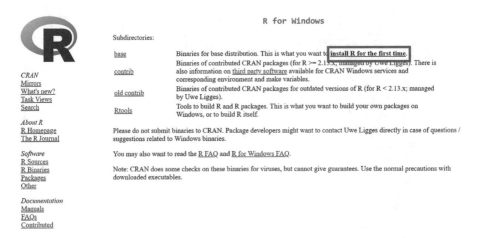

Figure 1-7. *Installing R on Windows*

On the next page, there will be a link to download the R installer. Click that to get the windows installer. It will be named "Download R for Windows". Click that to download the R installer. In a moment, you will find the R install executable file in your Downloads folder.

Double click the installer file and click through the prompts. Simply select the defaults and then let the installer finish.

Installing R on Linux

Each version of Linux has different set up instructions to follow to install R. Generally, you will use a package manager to install R from a command line. You can do something like this:

```
sudo apt-get update
sudo apt-get install r-base r-base-dev
```

However, each version of Linux might be a little bit different here, and you should double check the CRAN website to see any more detail on what you may need to do for your version of Linux. Click "Download R for Linux" link on https://cran.r-project.org and then choose your version of Linux to get extra guidance.

Install Git

Git is a popular version control system that is used to manage different versions of your code files. RStudio will integrate git into your workflow and help you visualize your work.

Install Git on Mac

To install git on Mac, go to this URL: https://git-scm.com/download/ mac and if your download does not start immediately, click the text "click here to download manually" to download the git installer. Double click the file that appears in your Downloads folder. In the folder that appears, click the pkg file to begin the installation process. You may get a security notification that looks like the dialog box pictured in Figure 1-8.

Figure 1-8. *Security warning on Mac*

You can override this security feature by going to your Mac's Security and Privacy setting's general tab and then selecting "Open Anyway" next to the file that you downloaded. See Figure 1-9 for an example.

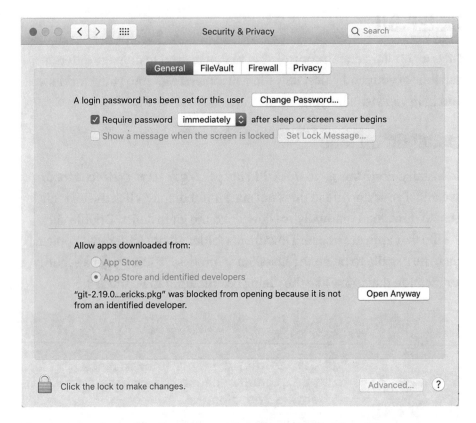

Figure 1-9. *Override Security settings to install git on Mac*

Now simply click through the installation screens until you get the notification that the installation is complete. You can use your Mac's Terminal app to verify that git is installed by typing git --version into the Terminal window.

Install Git on Windows

To install git on Windows, go to this URL: `https://git-scm.com/download/win` and if your download does not start immediately, click the text "click here to download manually" to download the git installer.

Double click the file that appears in your Downloads directory. Click Next for each screen that appears and simply choose the default settings. Once the installer has completed click Finish.

Verify that git was installed by going to the Windows Command Prompt and typing in "git –version". You should see a response that shows you the latest version of git.

Install Git on Linux

If you use Linux and you don't already have git installed, you can use your package manager to install git from the command line:

```
sudo apt install git-all
```

Here we used apt but note that you will have to use the package manager for your distribution.

Install RStudio

Now that we have git and R installed, we are ready to install RStudio itself.

Go to the RStudio downloads page: www.rstudio.com/products/rstudio/download/. You will be presented with a screen that shows all the available RStudio options. There are open source and commercial licenses available for both the desktop and the server versions of RStudio. We are interested in the open source (free) desktop version of RStudio. The other versions are geared toward businesses that need a commercial license and a high level of technical support. As shown in Figure 1-10.

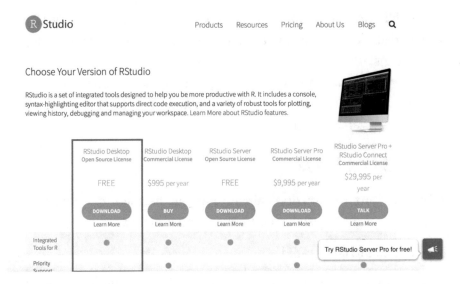

Figure 1-10. *RStudio download page*

Click the green pill that says Download and the screen will scroll down to a list of installers for Windows and Mac. Click the installer for your operating system and wait for the download to complete. Head over to your Downloads folder to find the installer; it will be named something like RStudio-1.1.463 with a dmg file extension for Mac and an exe file extension for Windows.

Double click the installer and follow the prompts. On Mac, a screen will pop up with an RStudio icon and a shortcut to your Applications directory. Simply drag the RStudio icon to the Applications folder shortcut.

On Windows, you can simply follow the prompts and choose the default locations. When the installation is complete, simply click the Finish button.

Verify RStudio

Before we move on, let's make sure that RStudio is installed and working. Search for the RStudio app icon on your system and double click to open

12

up the app. If you are having trouble finding the icon, go to the search icon in the upper right-hand area on your Mac screen or the lower left area on your Windows screen. Type in "RStudio" and you should see a shortcut to the RStudio app appear. Double click this icon.

If you are using Mac, a security popup will appear as shown in Figure 1-11.

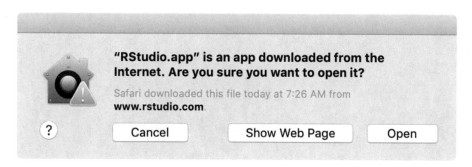

Figure 1-11. *RStudio security popup*

Click Open to head to RStudio. You should now see the RStudio app appear on your desktop. It will look something like this.

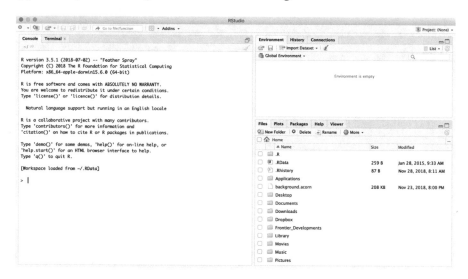

Figure 1-12. *RStudio IDE Initial Screen*

Congratulations, you now have RStudio IDE installed and ready to go!

CHAPTER 2

Hello World

Now that we all have RStudio setup, let's start using this program to help us become better R coders. The first thing we want to do is open up RStudio and start to work directly with R in the Console. This is our Hello World moment.

Interactive R Console

The first time you open RStudio you should see a window that takes up the entire left side of your screen. This is your Console screen and yours should look similar to what is featured in Figure 2-1.

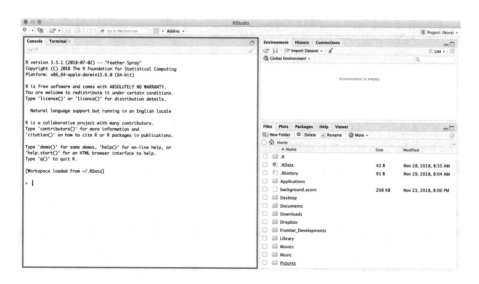

Figure 2-1. *Console screen*

© Matthew Campbell 2019
M. Campbell, *Learn RStudio IDE*, https://doi.org/10.1007/978-1-4842-4511-8_2

You should see information about the version of R that you are using. At the very bottom is a curser >. The space to the right of the curser is where you can type R commands directly into the Console.

Try typing in 1 + 1 and then hitting Enter. You should see [1] 2 appear on the line below. 2 is the answer to the question 1 + 1 and the 1 in parenthesis [1] is the position in the vector that corresponds to the answer 2.

Vectors

A central concept in R is the vector data structure. Vectors are lists of objects of the same data type. Many operations in R will return answers as a vector even if the question you ask, like 1 + 1, which returns only one value. [1] 2 is a vector of one numeric data type which has a value of 2 in the first position.

Hello World

Now let's see another type of vector in our Console by going through with the traditional Hello World example. Type in the following set of characters into your Console screen exactly as shown:

```
"Hello World"
```

Don't forget to include the quotation marks! Now press Enter and you will see another vector appear but this time it reads [1] "Hello World". Like before, this vector has one item in the first position, but the object this time is a string of characters showing the famous statement.

Let's try our Hello World a different way with the help of a function that we can use to create vectors with lengths of more than just one item. We can use the letter c and include a comma-separated list of objects to create a vector. All the objects in our vector must be of the same type (no mixing numbers and characters). If our objects are different types, R will simply convert them to the most general type that will accommodate the data that you supply. Type c("Hello", "World") into your Console window and then press Enter. This time you will see [1] "Hello" "World" print out to your Console.

While this appears very similar to what we saw before, it's a little bit different. This vector has two items, although the Console is only showing the [1] on the line. However, you can see that there are two objects above since each has its own set of quotes. R is simply saving space on the output screen by only showing item numbers of each row of output. Let's take a look at bigger vector to see how this works.

Type in 1:100 into your Console and press Enter. You should see something like the image displayed in Figure 2-2.

```
[Workspace loaded from ~/.RData]

> 1:100
  [1]   1   2   3   4   5   6   7   8   9  10  11  12  13  14  15  16  17  18  19  20
 [21]  21  22  23  24  25  26  27  28  29  30  31  32  33  34  35  36  37  38  39  40
 [41]  41  42  43  44  45  46  47  48  49  50  51  52  53  54  55  56  57  58  59  60
 [61]  61  62  63  64  65  66  67  68  69  70  71  72  73  74  75  76  77  78  79  80
 [81]  81  82  83  84  85  86  87  88  89  90  91  92  93  94  95  96  97  98  99 100
> |
```

Figure 2-2. *100 Item vector printout*

The : is a shortcut that you can use to define a range. Typing in 1:100 will give you every number between 1 and 100. The result is a vector and you can clearly see how each row will report back the item position of the data value in the first column.

Let's get back to our tour of RStudio IDE. We will get more about R programming and vectors as we get deeper into learning how to use the tools.

Terminal

The Console screen that we just introduced is what you would have available if you simply downloaded R and started coding. While this is great for interactively working with R, RStudio offers more features than a simple Console. Let's go through some of the RStudio features that make RStudio an integrated development environment.

Right next to the Console tab at the top of the Console screen is a tab that reads "Terminal". If you click that you will get another screen that looks similar to the Console but is a command line for your operating system. This is handy for situations that will come up when you need to work in this way.

This is a feature that will be welcome to command line ninjas and good to remember for the rest of us who at times need to jump down into the OS. Click the Terminal tab and type in a command that you might use. For example, you might want to see a list of files and folders so you could type ls on the your command line as we did in Figure 2-3.

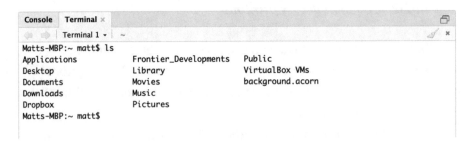

Figure 2-3. *Terminal command line*

The actual commands that you can use depends on your operating system. You can also access a list of helper actions if you use the dropdown menu that is right under the Terminal tab in Figure 2-4.

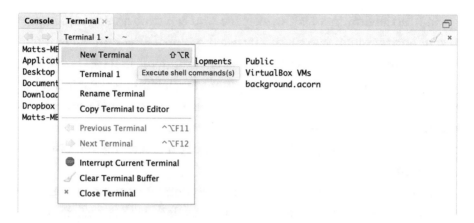

Figure 2-4. *Command shortcuts*

Environment

The RStudio environment tab is a handy feature located toward the right of the Console screen. There are three tabs: Environment, History, and Connections. The environment tab is pictured in Figure 2-5.

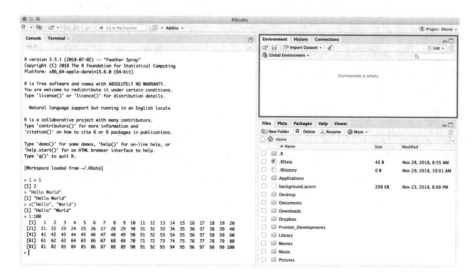

Figure 2-5. *Environment tab*

In R, an environment is a workspace. You can store objects and data in your environment. We will simply use the Global Environment in our examples, but you should be aware that environments in R can be a rich topic and are used in advanced R programming techniques. For instance, each R function (a unit of reusable code) will get its own environment and programmers can allocate their own environments for a variety of purposes. But most people don't need to worry much about the subtleties of environments in this way.

Right now, your environment should be completely empty because we have not yet created any objects or loaded any data. Let's change that by creating some variables. Click into your Console and create an object by assigning the "Hello World" string to an object named greeting.

```
greeting <- "Hello World"
```

Before we look at what happened in the environment window, we should talk about this new syntax I tried to sneak past you. Let's decompose the statement above starting the very right.

`"Hello World"` is the same string we type in at the beginning. A string is a set of characters enclosed in quotes. Like numbers, strings are a kind of data. The odd set of symbols to the left of the string `<-` is called the assignment operator. The assignment operator's job is to assign a data value to an object.

The object here is to the left of the assignment operator and it's labeled greeting. By typing in the code above and pressing Enter you assigned the value of `"Hello World"` to the object greeting and put that into the Global Environment. Now you can view the contents of the Global Environment in the right-hand area of the screen as pictured in Figure 2-6.

Figure 2-6. *Viewing objects in Environment tab*

This is really nice, as you are conducting a data analysis you will have visibility into the current state of your work. You can see the results at a glance and you can verify that the variables you expect are the right ones. This feature alone makes RStudio a must-have tool in your data programming kit. But there is more still.

The environment widget also has the ability to save your work and open up other environments. So, if you've been hacking on a data problem all day and want to save all our data and variables you can do that by using the floppy disk icon. To open an environment, simply use the folder icon and then navigate to an environment file that you have saved.

Importing Data

Another game changing feature which will help many people is the ability to import data from other data analysis tools like SAS and Excel. Click the Import Dataset button and you will see the options you have in import data in your analysis as pictured in Figure 2-7.

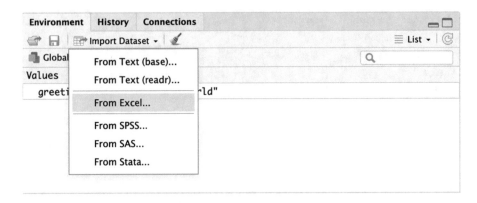

Figure 2-7. *Importing datasets*

Clicking through to these options will start up an import wizard that you can use to easily import data into RStudio from any of these sources. Finally, on this screen you also have a friendly broom icon that you can use to erase everything from your environment.

Datasets

To illustrate the import feature, we downloaded a dataset from a dataset used for the Food Pyramid from data.gov https://catalog.data.gov/dataset/mypyramid-food-raw-data-f9ed6. The Food Pyramid web site is shown in Figure 2-8.

This data is stored in Excel and we can simply use the Import Dataset Tool to import the dataset into R and inspect it. If you would like to follow along with this example, visit the URL above and download the zip file from the webpage.

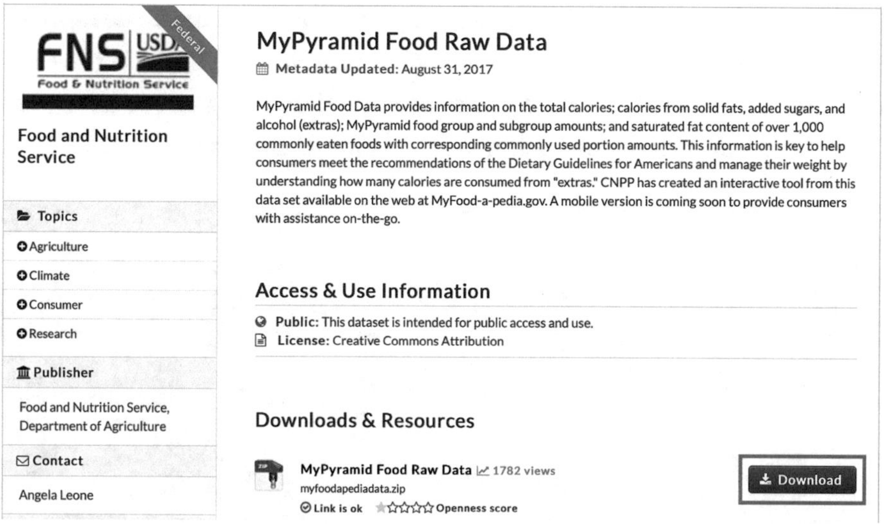

Figure 2-8. *Food pyramid data download site*

Open up the zip file and locate the file named Food_Display_Table.
xlsx. This is the dataset in Excel format that we will be looking at. Now in
RStudio, click the button "Import Dataset" and choose "From Excel...".
Click the "Browse" button. See Figure 2-9 for an example of this screen.

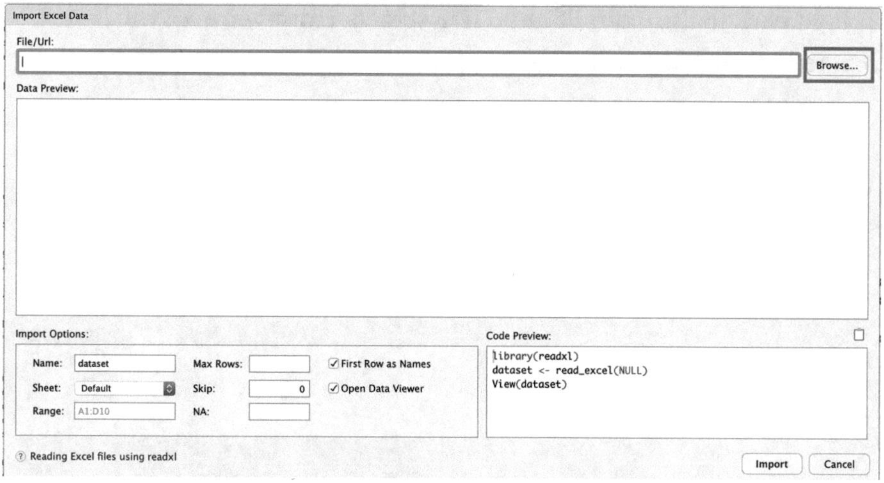

Figure 2-9. *Importing Excel data*

Next, navigate to the location where you located the Excel file named Food_Display_Table.xlsx. The screen will fill up with the first few rows of the data and also some code that can be copied that we can use later to re-import this dataset without using the interface as shown in Figure 2-10.

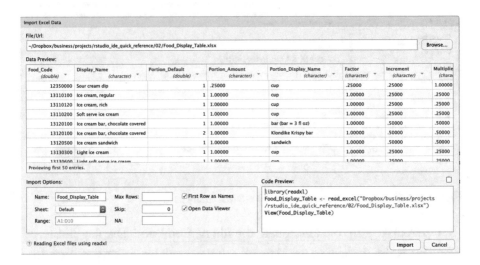

Figure 2-10. *Import data preview*

Click "Import" and the Import Dataset will disappear and you will be brought back to the main RStudio IDE screen. However, a lot will have changed as you can see in Figure 2-11.

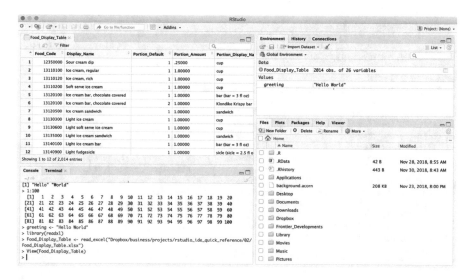

Figure 2-11. *Dataset opened in R*

You can see that the data has appeared on our screen. We can use the controls on this new screen to filter the data and to search the dataset. The dataset has also appeared in our Global Environment and has been automatically named Food_Display_Table. This is the name we would use to reference this dataset in R code. For instance, we can use the R head function in our Console to see the first few rows of data.

The output displayed in Figure 2-12 gives us even more information about this dataset. Once we learn more about data and R, we can use the name of the dataset as an input into our analysis.

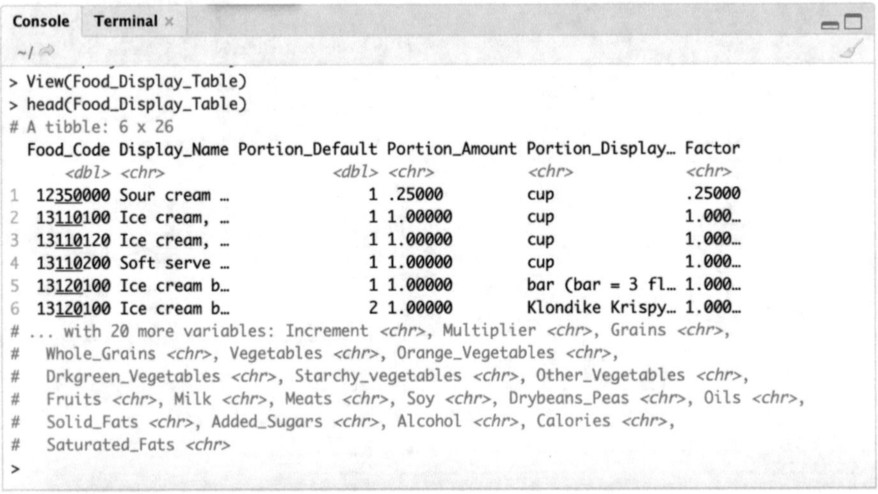

```
Console    Terminal ×                                                    ▬ ☐
~/ ⇨                                                                        ⟋
> View(Food_Display_Table)
> head(Food_Display_Table)
# A tibble: 6 x 26
  Food_Code Display_Name Portion_Default Portion_Amount Portion_Display… Factor
      <dbl> <chr>                  <dbl> <chr>          <chr>            <chr>
1  12350000 Sour cream …               1 .25000        cup              .25000
2  13110100 Ice cream, …              1 1.00000        cup              1.000…
3  13110120 Ice cream, …              1 1.00000        cup              1.000…
4  13110200 Soft serve …              1 1.00000        cup              1.000…
5  13120100 Ice cream b…              1 1.00000        bar (bar = 3 fl… 1.000…
6  13120100 Ice cream b…              2 1.00000        Klondike Krispy… 1.000…
# … with 20 more variables: Increment <chr>, Multiplier <chr>, Grains <chr>,
#   Whole_Grains <chr>, Vegetables <chr>, Orange_Vegetables <chr>,
#   Drkgreen_Vegetables <chr>, Starchy_vegetables <chr>, Other_Vegetables <chr>,
#   Fruits <chr>, Milk <chr>, Meats <chr>, Soy <chr>, Drybeans_Peas <chr>, Oils <chr>,
#   Solid_Fats <chr>, Added_Sugars <chr>, Alcohol <chr>, Calories <chr>,
#   Saturated_Fats <chr>
>
```

Figure 2-12. *R head function*

History

Click over to the next tab to see a log of all the commands that you entered into this R session so far. This is a big help when you have found result, but you can't quite remember how you got to a certain point. You can simply look at your log to see the list of steps you took.

The History tab shown in Figure 2-13 also has convenient functions that you can use to put a line of code that appears in your history back into the R Console. You can also search for a specific line of code using the text area with the search icon. This can be handy when you know that you have made a change to an object, but you are having trouble locating precisely where in the potentially long list of commands the changes were made.

Figure 2-13. *History tab*

Connections

Finally, the Connections tab gives you a way to connect to data sources such as local databases or a Spark cluster. These types of data storage solutions are used when datasets are too big to simply be stored locally on your computer. If you click into the Connections tab you will see an empty list, but there are options to connect to ODBC (databases) or Spark.

If you have connection information for a database that you need for your project, you can use this interface to connect. RStudio will guide you through the connection process and install any packages that you need to work with the data source. This window will also show any data source connections that you have created yourself with R code.

Conclusion

In this chapter, we introduced R vector, which is an essential R data structure. In addition, we have already been exposed to a good deal of the basic functionality of the RStudio IDE. Finally, we have a brief introduction to datasets in R. As we move on through our tour of the RStudio IDE, we will encounter these objects again and go into more detail, so you can start to see how this will help you with your data analysis.

CHAPTER 3

RStudio Views

Let's continue our tour of RStudio IDE. RStudio has a four-pane window layout and we have looked closely at two of the panes already in the last chapter. We have seen the Console screen, where you can write interactive R code, and we saw the Environment and History screens, where you can see the state of your R environment. Our data frame also opened up in a third pane, but we can do more with this part of the RStudio interface. Let's take a look at the remaining screens that we will be using now.

Files, Plots, Packages, Help, and Viewer Pane

By default, the pane on your lower right-hand side will contain tabs labeled Files, Plots, Packages, Help, and Viewer. The Files tab will show you the files in the current working folder. The working folder is the folder that R will save your work to. This pane essentially is your Windows Explorer or Mac Finder app integrated into RStudio.

Files

The Files tab includes convenient functions that you can use to open and generally work with files without needing to leave RStudio. This includes tasks like deleting and renaming files. You can also set the R working folder as you can see in Figure 3-1.

© Matthew Campbell 2019
M. Campbell, *Learn RStudio IDE*, https://doi.org/10.1007/978-1-4842-4511-8_3

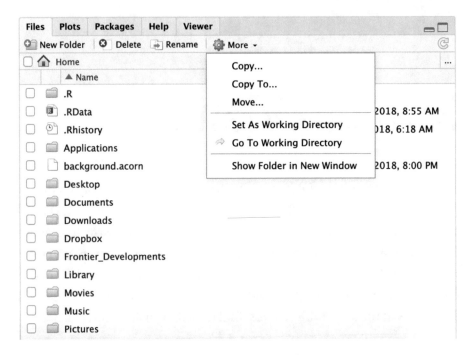

Figure 3-1. *Files tab*

Plots

You use the Plots tab to view the plots that you create during your R analysis. If you click over to this tab, you will see an empty screen with a few function buttons, which is pictured in Figure 3-2.

Figure 3-2. *Plot view functions*

You can view plots, export plots to files, and also navigate sets of plots that you created. This widget helps you stay organized as you go through and investigate your data. To demonstrate this feature more, we will need to hop over to the Console screen, do some data munging, and create a plot.

Introduction to Data Munging

We can use the data that we imported in the previous chapter. If you have been following along, that data is named Food_Display_Table. Before we use this data to make a plot, we are going to have to do some "data munging". Data munging is the process of restructuring the data to make it work for your analysis.

Note   We are using the terms "dataset" and "data frame" interchangeably above; however, the technical name for the object that we are working with in R is "data frame". The term "dataset" is a generic term that applies to any set of data in any analysis environment.

For this plot, we want to look at two columns in the Food_Display_ Table data frame: Calories and Portion_Amount. It seems like these two items may have a relationship, so it makes sense to see if using a plot will help us see any patterns that might act as a clue to help us understand what this data is telling us.

The first thing I want to do is check data type of the columns that I am interested in. You can use the R class function to do this:

```
class(Food_Display_Table$Calories)
```

You have the name of the function, class, and in the parenthesis, you include the object that you want to inspect. We already saw the data frame name (Food_Display_Table) but now we also have a dollar sign $ and the name of the variable Calories. The dollar sign in R is used in data frames to reference a column. Each column in the data frame Food_Display_ Table can be referenced in this way. Above we used the class function to inspect the datatype of the Calories column. It returned "character".

For our plot, we need the data to be numeric. Luckily, it's super easy to change; make a column of data numeric by using the as.numeric function and then assigning the results back to the data frame:

```
Food_Display_Table$Calories <- as.numeric(Food_Display_
Table$Calories)
```

This code replaces the character version of the `Calories` data with a numeric version of the same data. You can double check the data type by using the same class function like this:

```
class(Food_Display_Table$Calories)
```

This time the Console should return "numeric". Before moving on to the plot, we still need to do this for the `Portion_Amount` column. Follow the same process above as we used for `Calories`. You should see something like Figure 3-3 in your Console screen once everything is complete.

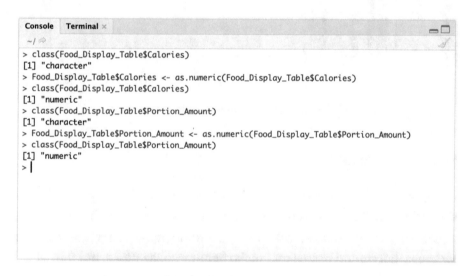

```
> class(Food_Display_Table$Calories)
[1] "character"
> Food_Display_Table$Calories <- as.numeric(Food_Display_Table$Calories)
> class(Food_Display_Table$Calories)
[1] "numeric"
> class(Food_Display_Table$Portion_Amount)
[1] "character"
> Food_Display_Table$Portion_Amount <- as.numeric(Food_Display_Table$Portion_Amount)
> class(Food_Display_Table$Portion_Amount)
[1] "numeric"
>
```

Figure 3-3. *Changing data types*

Plot

Now we are ready to make a plot. Type this code into your Console window:

```
plot(Food_Display_Table$Calories, Food_Display_Table$Portion_
Amount)
```

Now a plot in your Plots tab will appear that looks like Figure 3-4.

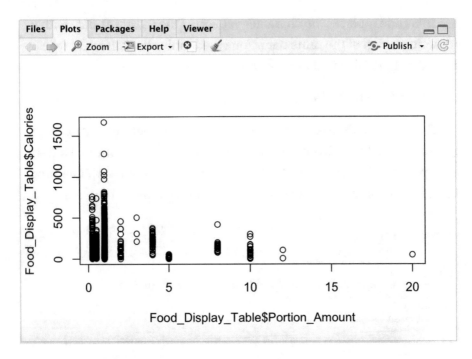

Figure 3-4. *Plot in R*

Figure 3-4 shows Calories on the Y axis and `Portion_Amount` on the X axis of the plot. This is a pretty basic plot, but it's good enough to hint at a relationship between these data points. At some point, you will learn about packages that extend base R and make better visualizations, but they all essentially will work the same way in the Plots tab.

Packages

The next tab labeled "Packages" will show you a list of R packages that you have available. See Figure 3-5 to see what the Packages list looks like.

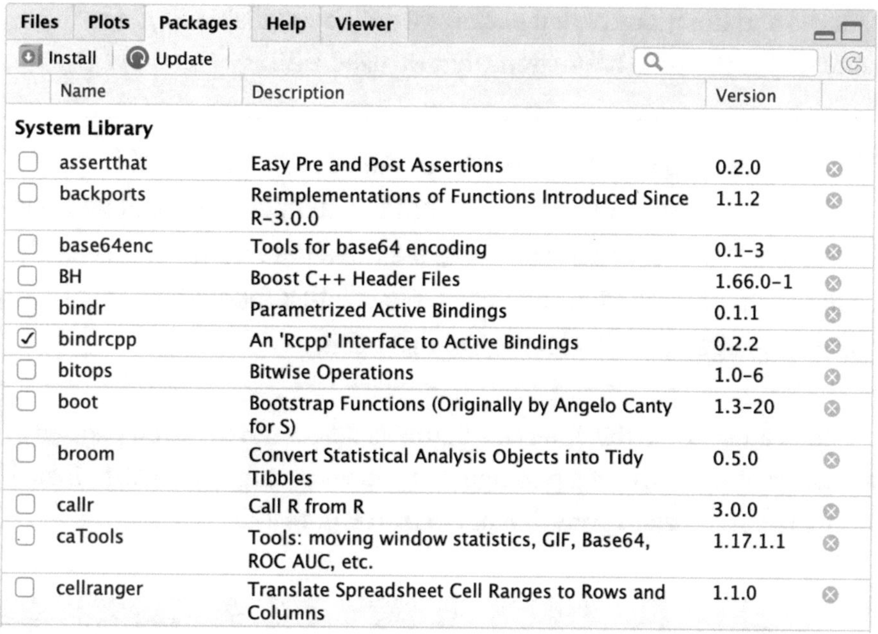

	Files	Plots	Packages	Help	Viewer		▬ ☐

	Name	Description	Version	
System Library				
☐	assertthat	Easy Pre and Post Assertions	0.2.0	⊗
☐	backports	Reimplementations of Functions Introduced Since R-3.0.0	1.1.2	⊗
☐	base64enc	Tools for base64 encoding	0.1-3	⊗
☐	BH	Boost C++ Header Files	1.66.0-1	⊗
☐	bindr	Parametrized Active Bindings	0.1.1	⊗
☑	bindrcpp	An 'Rcpp' Interface to Active Bindings	0.2.2	⊗
☐	bitops	Bitwise Operations	1.0-6	⊗
☐	boot	Bootstrap Functions (Originally by Angelo Canty for S)	1.3-20	⊗
☐	broom	Convert Statistical Analysis Objects into Tidy Tibbles	0.5.0	⊗
☐	callr	Call R from R	3.0.0	⊗
☐	caTools	Tools: moving window statistics, GIF, Base64, ROC AUC, etc.	1.17.1.1	⊗
☐	cellranger	Translate Spreadsheet Cell Ranges to Rows and Columns	1.1.0	⊗

Figure 3-5. *Package viewer*

An R package is an extension to the base R system you get when you installed R in Chapter 1. R packages are written by community contributors and they are available for anyone to use. Packages are also managed by the CRAN network and you can manage all of this from your Packages tab.

What you see in the screenshot in Figure 3-5 is a list of all the packages that are installed on your computer. The checkmarks indicate that the package is loaded into your R environment and is ready to use. If you check a package, code will appear in your Console screen that loads the package.

For example, to use the broom package we would simply click the checkbox and in the Console screen this would appear:

```
library("broom", lib.loc="/Library/Frameworks/R.framework/
Versions/3.5/Resources/library")
```

The exact location would depend on the version of R and the operating system that you are using. Once a package is loaded in this way, all the datasets and functions of that package would be available to you in the same way as the base R functions that we used earlier.

Note   The broom package is part of the tidyverse family of R packages. Broom is used to transform the complex output of analysis functions like linear regression into a simple data frame that is easily used as an input to other functions. We will talk more about must-have packages like this later on in this book.

You can also use this interface in the Packages tab to install packages. If you know of the name of a package that you want simply click the "Install" button to get the dialog box that appears in Figure 3-6.

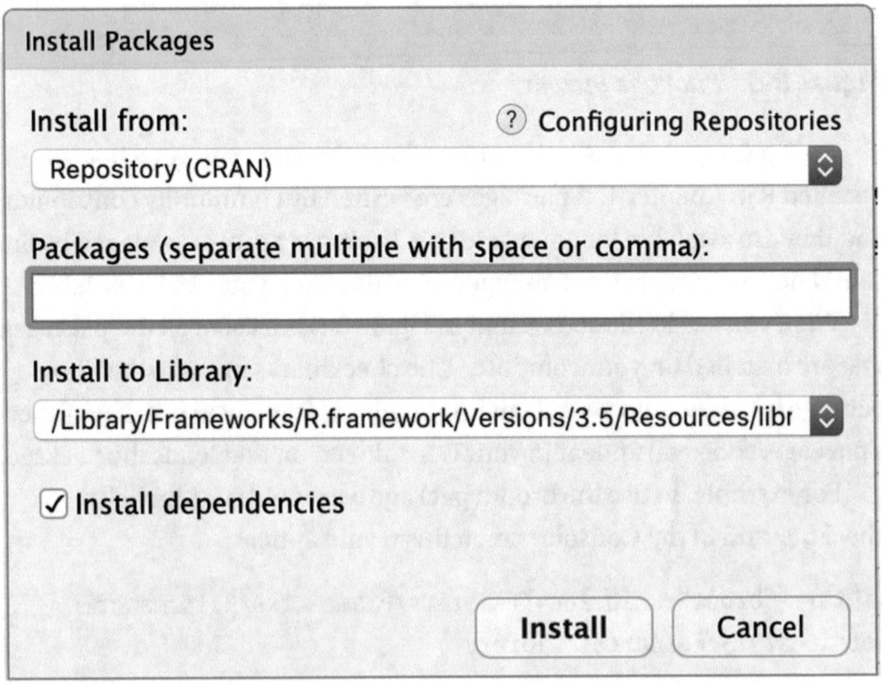

Figure 3-6. R Packages Tab

Then type in the name of the package that you want and click the "Install" button. Leave the defaults in place since this gives you all the packages that are available on the CRAN network.

Finally, each package name that appears is a hyperlink that you can use to get to the documentation for the package. This documentation gets installed when you install the package. Simply click the package name to learn about how to use the package. This will automatically take you to the Help widget that appears to the right of your Packages tab.

Help

The Help tab is set up like a web browser and includes elements to help you search for topics. You can navigate through the content that appears as if it was a web page as pictured in Figure 3-7.

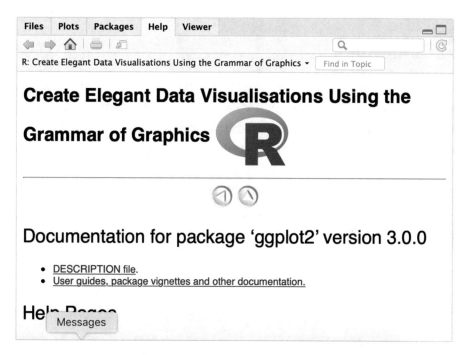

Figure 3-7. *Help tab*

In the screen pictured in Figure 3-7, you can see the documentation for the `tidyverse` package called `ggplot2`. This package will help you create richer data visualizations than you could just by using the base plot function. You would use the Help screen to learn more about packages like this or to get help with specific functions by using the search feature.

Viewer

The final tab in the lower right-hand year is the Viewer tab. This is an area where output will appear from some of the more advanced features of RStudio reporting such as R markdown documents. R markdown documents are used to present code and results together in a reproducible way that is used to share results. We will address R markdown later on in this book.

Conclusion

By this point, you have a pretty solid foundation for the basics of working with RStudio IDE. You should understand how to navigate through all the panes and interfaces you need to conduct a quick analysis using all the features in the four panels that are available in the default state in RStudio.

Along the way, we also went into how to do some data munging, data exploration, and data visualization with base R.

Next, we will continue to explore how RStudio can help us keep organized and build analysis that is clear, insightful, and reproducible.

CHAPTER 4

RStudio Projects

What we have learned so far adds a great deal of integration to R programming. But we are missing a way to bring all of that together in a format that we can reuse. This is where RStudio projects come into play.

RStudio projects are used to organize all the code, reports, and any other assets used in an analysis. When we create a project, RStudio will provide the needed files and everything else needed for your type of project.

Create a New RStudio Project

Create a new RStudio project by selecting "File" and then "New Project..." from the menu bar at the top of your RStudio screen as pictured in Figure 4-1.

© Matthew Campbell 2019
M. Campbell, *Learn RStudio IDE*, https://doi.org/10.1007/978-1-4842-4511-8_4

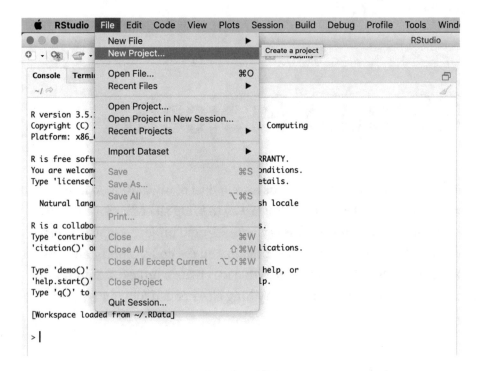

Figure 4-1. *Creating a new RStudio project*

Now a dialog box as pictured in Figure 4-2 will appear that presents a few options to create your new RStudio project.

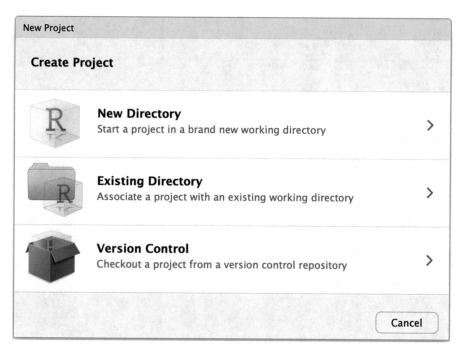

Figure 4-2. *RStudio project options*

Usually, you will simply choose the first option, "New Directory", which will create a new folder that you can name. All your configuration files will appear in this new folder. However, if you already have some code files that you want to start with, you may choose the second option, "Existing Directory". Finally, if you already have a version control system like git set up for a project that already exists, then you can choose the third option, "Version Control".

Project Types

For this example, we clicked on "New Directory" to create a fresh new project. The screen that appears will show you a list of RStudio project types as pictured in Figure 4-3.

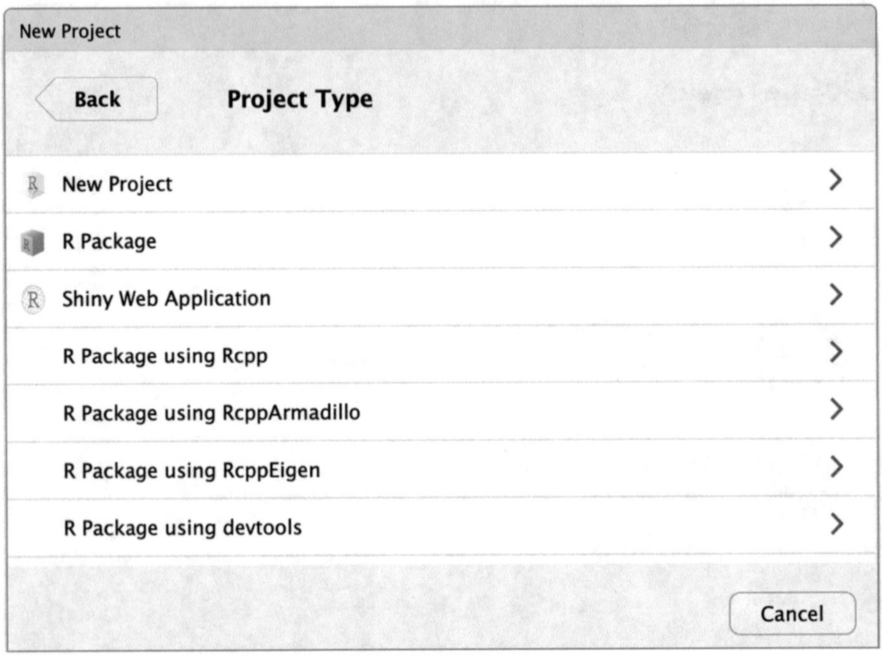

Figure 4-3. *RStudio project types*

The first three project types are the ones that you will use most often. The projects after the first three all include specialized code used for special use cases. Projects in "Rcpp" in the description all are used to integrate with C++ code libraries. The projects with "devtools" are used to create R packages with the devtools library and the sparklyr project will create a package that will work with Spark.

"Package" projects are set up to create a set of reusable R objects. This is a bit more advanced than what we are doing in this chapter and we will come back to this later. Just be aware that you can create RStudio project packages when you have code that you believe can be used across your projects.

For this example, choose "New Project" as pictured in Figure 4-4 to create project that will work best for an analysis.

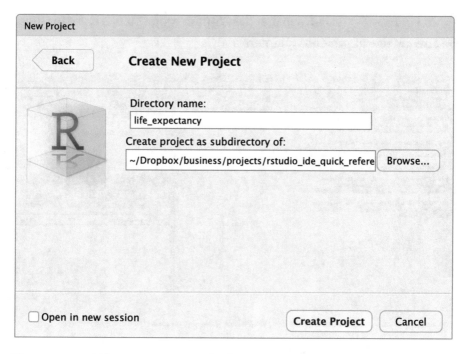

Figure 4-4. *Life expectancy RStudio project*

We named the project "life_expectancy" because we are going to do some data exploration from the data.gov site that shows life expectancy rates. For now, choose the directory (known as folder on Mac) where you want the project to be saved and then click the "Create Project" button to create this RStudio project.

RStudio Project Tour

Now we have a new space for our project. This new RStudio project will be stored in the folder that you typed into the dialog box pictured in Figure 4-4. RStudio projects get their own environments so if you are switching projects your work will be saved and organized. This is true for code files and anything else that you may need for an analysis. Let's take a look at what was set up for us.

In Figure 4-5, you can see that the RStudio project looks close to what we already saw in previous chapters.

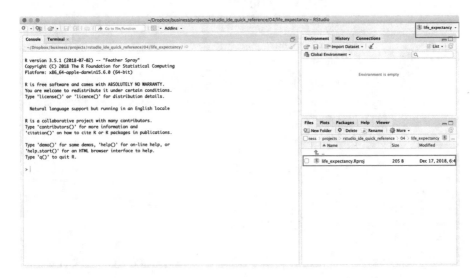

Figure 4-5. *New project screen*

However, in Figure 4-5, you can see two changes from the normal starting screen. A new file appears in the bottom right screen with your project name and the "Rproj" file extension. This file contains all the settings for your project. If you look closely in your file explorer, you can see that this file sits in the directory that you choose under sub-directory with the same name as your project.

If you click on your project name in the upper right area of your screen, you will find a list of helpful shortcuts as pictured in Figure 4-6.

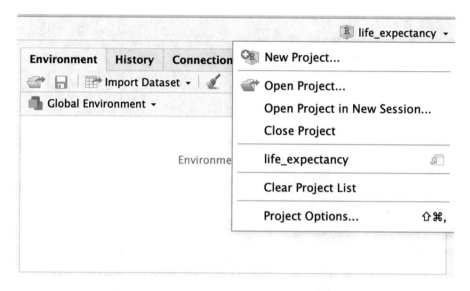

Figure 4-6. *Project shortcuts*

You can open and close projects, quickly navigate to projects that have been opened recently, and set RStudio options specifically for this project.

Project options are saved in the file with the extension "Rproj" in the folder that was created for your project. You can either click the file to open your project options or choose the option in the shortcut list pictured in Figure 4-6. When you do this, you will see a screen like Figure 4-7 that lists the options that you can set for this project.

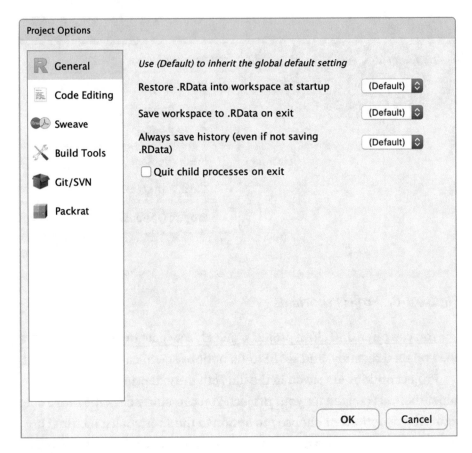

Figure 4-7. *Project options*

These project options are inherited from the global options that are set for RStudio and for the most part you can simply leave these alone. The first two screens are the ones that are most immediately useful. The "General" tab gives you options when it comes to saving the state of your analysis. The default is to save both your data environment and R history. Sometimes if your project needs to load massive datasets, you might want to turn these options down because it would make your project take longer to open. Generally, it's best to keep the defaults in place.

The next tab is "Code Editing" and as you can see in Figure 4-8; this lets you control how your code editor works.

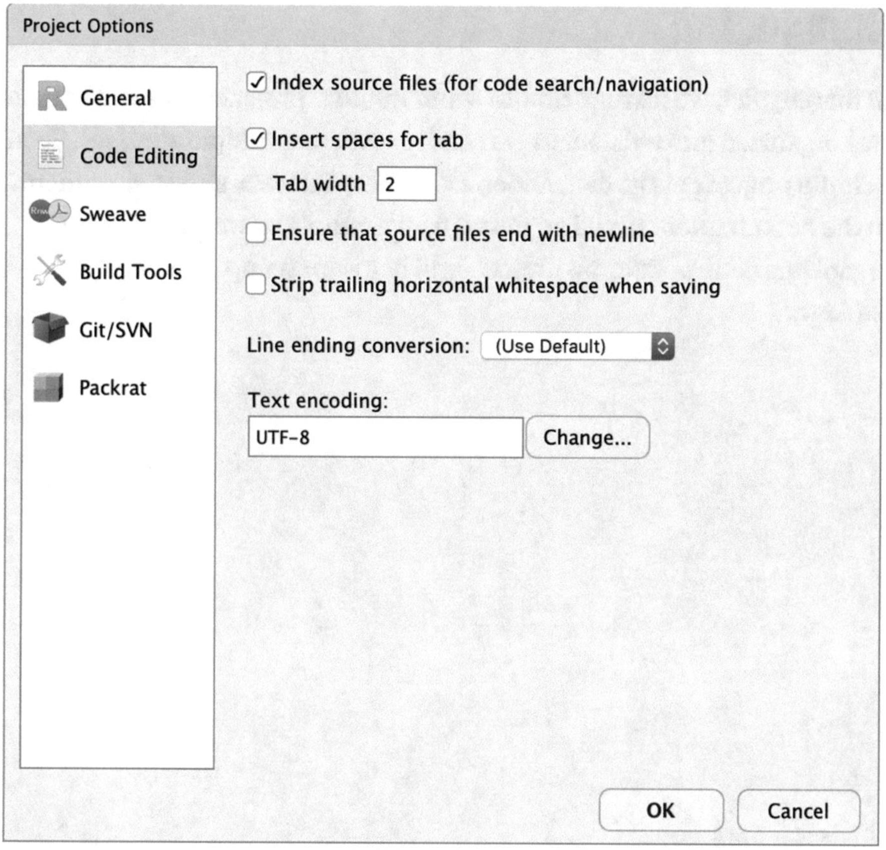

Figure 4-8. *Code editing options*

You will want to keep these default settings generally. Make sure that the first option "Index source files" remains checked because that will ensure that some of the code tools we will talk about later will work correctly.

The remaining four options tabs contain settings for features that we will revisit later on in this book. These options generally are used for reporting, building packages and other specialized features that we not quite ready to integrate into our workflow.

Conclusion

In this chapter, we learned how to setup RStudio projects. This will help us stay organized and efficient as we start to work on multiple analyses. Each RStudio project gets its own folder, set of files, datasets, and environment. In the next chapter, we will go through an example so we can see how organizing code in RStudio projects will help you set up an efficient workflow.

CHAPTER 5

Repeatable Analysis

Let's take a look at how to create a repeatable analysis using an RStudio project. For this example, we will be working with the project we created in the previous chapter labeled "life_expectancy". The dataset will be from the National Center for Health Statistics and will describe life expectancy since 1900 in the United States.

If you have been following along, you will already have this project set up in RStudio. If not, follow the steps in the previous chapter to create a new RStudio labeled "life_expectancy".

Organizing Datasets

One thing that we will want to do is read the data into the project. To stay organized, it helps to create a folder to store your data. Use the file viewer to create a new folder as shown in Figure 5-1.

© Matthew Campbell 2019
M. Campbell, *Learn RStudio IDE*, https://doi.org/10.1007/978-1-4842-4511-8_5

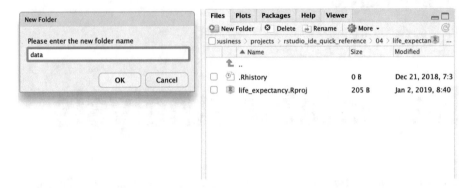

Figure 5-1. *Create a new folder*

Go to Files and then "New Folder" type in the word "data" and then click "Ok". You will now have a new folder that is associated with this project.

Download Life Expectancy Data

Next, follow this link to go to the page where the dataset is located:

```
https://catalog.data.gov/dataset/age-adjusted-death-rates-and-
life-expectancy-at-birth-all-races-both-sexes-united-sta-1900
```

This page will include a detailed description of the data. We want to download the csv file into the "data" folder that we just created. Scroll down to the area of the page pictured in Figure 5-2.

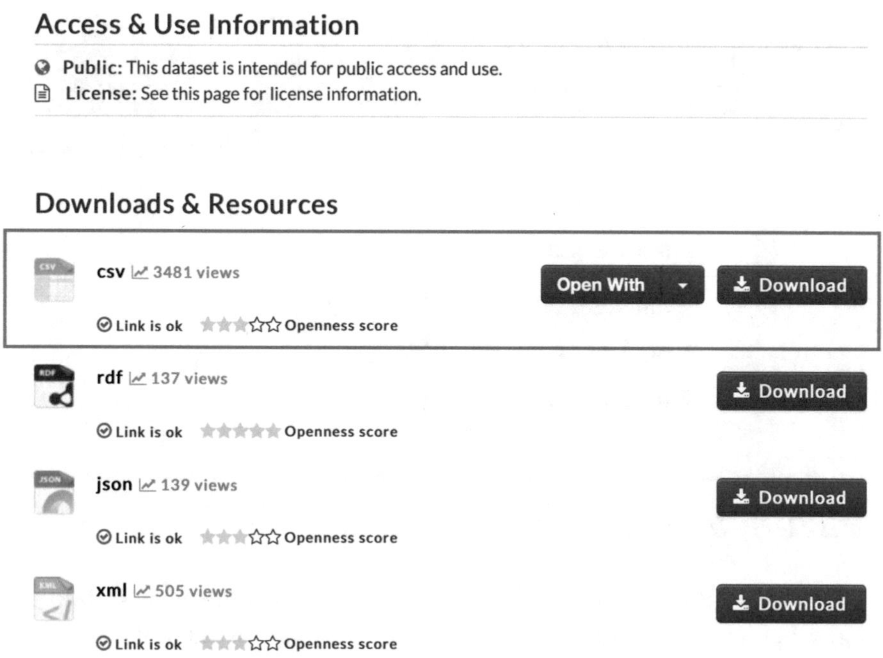

Access & Use Information

🌐 **Public:** This dataset is intended for public access and use.
📄 **License:** See this page for license information.

Downloads & Resources

csv 📈 3481 views
⊘ Link is ok   ★★★☆☆ Openness score
[Open With ▾] [⬇ Download]

rdf 📈 137 views
⊘ Link is ok   ★★★★★ Openness score
[⬇ Download]

json 📈 139 views
⊘ Link is ok   ★★★☆☆ Openness score
[⬇ Download]

xml 📈 505 views
⊘ Link is ok   ★★★☆☆ Openness score
[⬇ Download]

Figure 5-2. *Life expectancy CSV download link*

Click the "Download" button. The csv file will appear in your downloads folder. Move the csv file from your downloads folder to the data folder that you created in your project. Note that you will have to use your computer's Finder or Explorer app to move this file.

Add Dataset to R Project

While we have the raw csv file saved in the data folder, we need our data to be in the dataframe format in order to be used in our analysis. The easiest way to read in the data in this format is to use the "Import Dataset" feature located in the environment view as pictured in Figure 5-3.

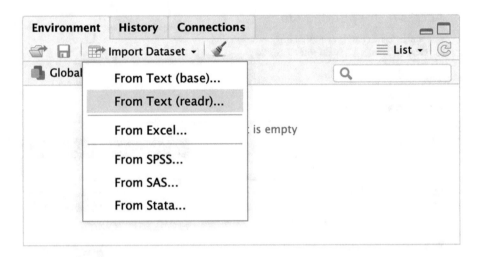

Figure 5-3. *Import dataset*

Go to the "Environment" view, click on "Import Dataset" and then choose "From Text (readr)...". A screen similar to Figure 5-4 will appear.

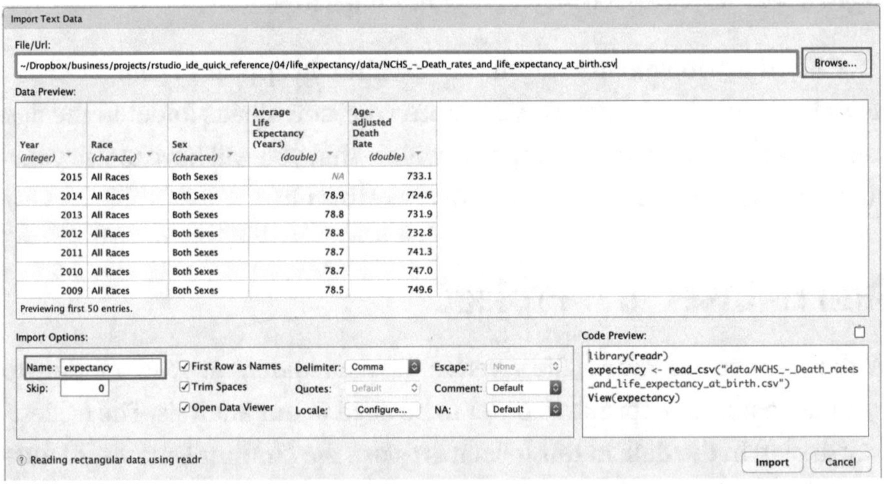

Figure 5-4. *Data import widget*

Use the "Browse" button to navigate to the csv file. This is highlighted in Figure 5-4 at the upper right-hand area of the screen. The default name of the dataframe will be the same as the file, but since this would have been very long you can change it right from this widget screen as shown in the highlighted area next to "Name" in the lower left-hand area of the widget. We simply named the dataframe "expectancy".

Once you have completed this process, the dataframe will appear in your R Studio project as pictured in Figure 5-5.

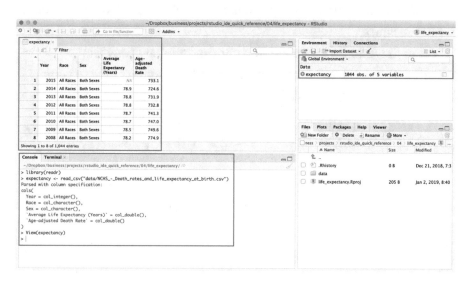

Figure 5-5. *Dataframe in R Studio project*

In Figure 5-5, you can see a sample of the dataframe at the top left. The code that was used to read in the data will appear in the Console at the bottom left and this code will also be retained in the history view. Finally, you will see an object with the name "expectancy" located in the "Environment" window at the top right. You can click this object to make the data appear again if you close the window at some point.

This is the first step of setting up your project. Make sure to save your project's state if you close RStudio. Otherwise you will need to repeat this process again if you want to work with your dataframe.

R Code Files

While it can be easy at first to use the provided widgets and interactive Console screen, if we want to redo our analysis or ask a colleague to run our analysis it would be difficult to retrace our steps. Our analysis is not really reproducible.

To make our work reproducible, we can add R script files to our project. What we are going to do now is add an R script file and then add the code that was automatically generated to this file so that we can reuse this in the figure.

Go to the RStudio menu bar and choose "File" ➤ "New File" ➤ "R Script". A new tab as pictured in Figure 5-6 will appear next to your dataframe labeled "Untitled1".

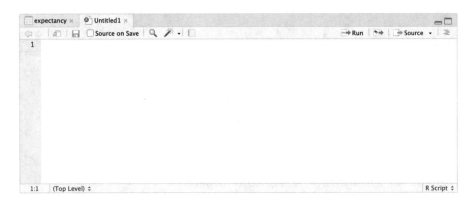

Figure 5-6. *R Script editor*

This script editor comes with very useful built in features along with a bar of visual shortcuts at the top. First, we will click the floppy disk icon to save this file into our project with a memorable name. Click the floppy disk icon and in the dialog box that appears enter in the filename "process_ data.R". The file will now appear in your file explorer and you can open the file up at any time by clicking on its name.

We will look at the other features here when we talk about code tools but for now be aware that we can save our work here and also run code using the "Run" and "Source" buttons above. For now, we want to retrieve the code that was automatically generated that read in the csv file and then do a little housekeeping on this dataframe.

Open up the "History" tab at the right and you should see the lines of code that were used to read in the csv file as pictured in Figure 5-7. Command-click to highlight the first two lines of code.

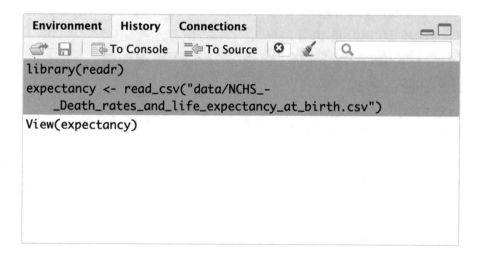

Figure 5-7.   *Retrieve code from History tab*

Click the "To Source" button to move this code into your R script file. Then click the floppy disk icon to save the file as pictured in Figure 5-8.

Figure 5-8.   *R Script file with retrieved code*

We can now rerun this step whenever we need to by clicking the "Source" button at the top. Also, we could send this project to a colleague and she could recreate our work in the same way so our analysis is on track to staying reproducible.

Before we move on we should add some comments to our code and clean up the dataframe so that it is easier to work with later on. The code below is one example of how this can be done:

```
# Import NCHS Dataset
# File download from:
#   https://catalog.data.gov
# Contains life expectancy data from 1900 in the US

library(readr)
expectancy <- read_csv("data/NCHS_-_Death_rates_and_life_
expectancy_at_birth.csv")

# Clean up dataframe
# Use shorter column names

names(expectancy)[1] <- "year"
names(expectancy)[2] <- "race"
names(expectancy)[3] <- "sex"
names(expectancy)[4] <- "life_expectancy"
names(expectancy)[5] <- "death_rate"
```

The next after the # are comments and the last five lines of R code replace the column names with shorter names that will be easier to work with. Don't forget to click the floppy disk icon to save the code in this file.

Data Exploration

One of the things you may want to do when you are starting a new project is to do some exploratory data analysis. We can inspect the dataset visually already, but it would be nice to have a more complete picture. We also may want to start to visualize the more important data points.

Summarizing Dataframes

Let's create a new R script file by going to the RStudio menu bar and choosing "File" ➤ "New File" ➤ "R Script". Just as before, a new file will appear. Click the floppy disk icon to save this file with the filename "data_exploration.R".

Now we want to see a summary of the expectancy dataframe, so we can use the summary function with the dataframe as an argument.

```
summary(expectancy)
```

Type this function right into your R script file and then click the "Run" button to code this code. You will see output that describes each variable (or column of the dataframe) as pictured in Figure 5-9.

```
> summary(expectancy)
      year            race               sex           life_expectancy  death_rate
 Min.   :1900   Length:1044        Length:1044        Min.   :29.10    Min.   : 616.7
 1st Qu.:1929   Class :character   Class :character   1st Qu.:56.60    1st Qu.:1040.2
 Median :1958   Mode  :character   Mode  :character   Median :66.60    Median :1541.4
 Mean   :1958                                         Mean   :64.12    Mean   :1614.1
 3rd Qu.:1986                                         3rd Qu.:73.60    3rd Qu.:2073.0
 Max.   :2015                                         Max.   :81.40    Max.   :3845.7
                                                      NA's   :9
```

Figure 5-9. *Dataframe summary*

The information in Figure 5-9 shows us a lot about this dataset. We can see that what type of data each column contains and we have some information about the numeric data. For instance, we know that the dataset covers the years 1900 to 2015 and that some years have a life expectancy of 29.10 while others had 81.4. We can also see that we have two character types race and sex, but we have little information about these. These are categories and will reveal how this dataset is organized.

To look at the race and sex categories we can use the R unique function and pass the column of data as an argument like this:

```
unique(expectancy$race)
```

If you hit the "Run" button, you will see this output:

```
[1] "All Races" "Black"      "White"
```

When you do the same for the sex category, you can see that we also have three categories available: "Both Sexes", "Female", and "Male".

We can tell from all this that our data will be organized by three categories: year, race, and sex. Also, most likely each year will have rows for all available combinations of race and sex. You can use the dataframe view to verify this.

Navigate to the first tab in your project to view the expectancy dataframe again. Click the filter button to see a subset of the data. Use the slider in the year column to only view rows that are identified by the year 1900 as pictured in Figure 5-10.

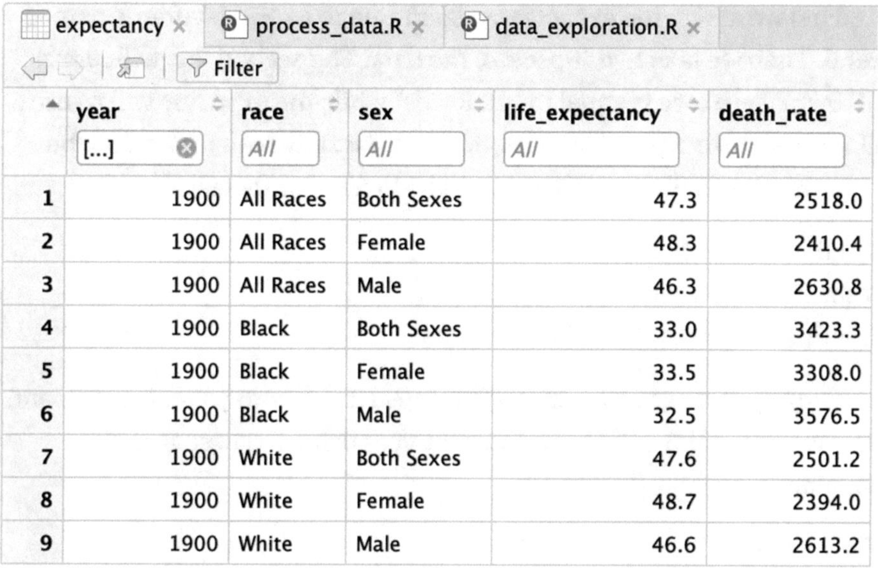

	year	race	sex	life_expectancy	death_rate
	[...]	All	All	All	All
1	1900	All Races	Both Sexes	47.3	2518.0
2	1900	All Races	Female	48.3	2410.4
3	1900	All Races	Male	46.3	2630.8
4	1900	Black	Both Sexes	33.0	3423.3
5	1900	Black	Female	33.5	3308.0
6	1900	Black	Male	32.5	3576.5
7	1900	White	Both Sexes	47.6	2501.2
8	1900	White	Female	48.7	2394.0
9	1900	White	Male	46.6	2613.2

Figure 5-10. *Filtered dataframe*

We will want to be careful in how we handle this dataframe now that we know how things are organized. For example, we will not want to simply aggregate datapoints by year since it contains overlapping categories and we will want to filter on these categories as we develop research questions.

You may also want to do some preliminary visualizations to start to get some insight into the numeric variables we are looking at. For instance, you may guess that life expectancy would have increased since 1900. To test that theory, you can look at only the rows in the dataframe with the race and sex category values of "All Races" and "Both Sexes".

```
trend <- expectancy
trend <- trend[trend$race == "All Races", ]
trend <- trend[trend$sex == "Both Sexes", ]
```

First we assign the expectancy dataframe to a new dataframe named trend. Then we take two subsets of the data. The second line will return all rows where race is equal to "All Races" while the third line will return all rows where sex is equal to "Both Sexes". Next, we can remove all the variables expect the ones that we will be looking at specifically.

```
trend$race <- NULL
trend$sex <- NULL
trend$death_rate <- NULL
```

Setting objects to NULL in R will remove them from your environment. Now we can plot the life expectancy by year using the base R plot function.

```
plot(trend)
```

This will automatically plot the two variables and your plot viewer will have the plot displayed for you as pictured in Figure 5-11.

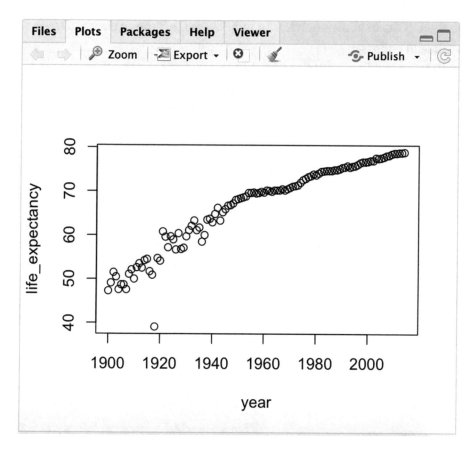

Figure 5-11. *Life expectancy by year*

As you can see, this plot supports our initial guess that life expectancy would increase over the years. The plot also shows that the observations in earlier years are not as consistent as observations in later years.

Conclusion

This chapter presents an example of how you might use RStudio projects to organize and make your work reproducible. RStudio projects make it easy to retrace your steps and build on your earlier work as you flesh out your analysis. In addition, RStudio projects make it much easier to share work with colleagues.

When you close RStudio after working on a project like this make sure to click "Save" when the IDE prompts you as you exit. This will save your work and history and make it easier to pick up where you left off in your project.

CHAPTER 6

Essential R Packages: Tidyverse

Now that you have the essential RStudio workflow, we are ready to start highlighting some of the tools that will really supercharge your data analysis. Not all of these tools are exclusive to RStudio and you may see them in other R environments, but they do fit in very nicely with the RStudio IDE.

The first order of business here is to introduce you to the must-have packages what will provide the cornerstone of your R toolkit. Many of these packages will simplify functions that already exist in R, while others will add functionality to the base system. RStudio along with these packages are what transforms R into a professional tool for data practitioners.

R Packages

An R Package is something that you can plug into RStudio to extend the basic functionality that is built in with R. One of the reasons that R has become so popular is because it has this rich ecosystem of packages that really make R a comprehensive platform for data science.

You can use RStudio to load, browse, install and remove packages or you can manage R packages using R code. RStudio's package support is discussed in detail in Chapter 3.

© Matthew Campbell 2019
M. Campbell, *Learn RStudio IDE*, https://doi.org/10.1007/978-1-4842-4511-8_6

Tidyverse

The packages that we will discuss are actually a collection of packages called "tidyverse". Tidyverse is mainly developed by Hadley Wickham who also works on RStudio. The tidyverse website, https://www. tidyverse.org, describes this set of packages best, *"The tidyverse is an opinionated collection of R packages designed for data science. All packages share an underlying design philosophy, grammar, and data structures."*

Note that the authors describe tidyverse as "opinionated." This means that tidyverse is more than a set of functions; it is also a prescribed way of doing data science. There is a tidyverse way of doing things. This is what they mean when they say tidyverse is "opinionated."

One of the things that tidyverse does well is smooth over base R's rough edges. tidyverse provides consistent wrappers over functions that already exist in base R that are easier to use and easier to remember. These packages also add more functionality to what is included in base R. For instance, ggplot2 (a tidyverse package) extends what you can do with base R's plot function but adds more plot types. Let's get into the core tidyverse packages.

To follow along with the examples below, you must install tidyverse and then import the library into your project. You may use the instructions on packages in chapter three to do this or you can type in this code to your R Console.

```
install.packages("tidyverse")
library(tidyverse)
```

magrittr

The magrittr package changes the flow of R coding by introducing the pipe operator %<%. The %<% operator pipes left-hand values into right-hand expressions. This allows us to create "pipelines" where we can immediately and clearly see what a block of code is doing.

In a programming language like R that depends so much on functions you will start to end up with code that is difficult to read like this:

```
round(mean(subset(expectancy, race == "All Races")$death_rate), 1)
```

This one line of code will return the average death rate in our expectancy dataframe we worked with in the last chapter. However, it is performing three tasks: filtering the dataset, calculating the average score, and then rounding the result. It is very hard to read code like this and as problems become more complex, code like this gets harder to read.

The magrittr package pipe operator %>% will help us tidy our code by removing the need to nest function calls. Here is an alternative way to write this code:

```
expectancy %>%
  subset(race == "All Races") %>%
  .$death_rate %>%
  mean() %>%
  round(1)
```

This code is a little bit longer, but it is also much clearer. You read it top to bottom and each %>% takes the object above it and sends it to the next function as the first argument. At the top we use %>% to send the expectancy dataframe to subset as its right argument.

Now we can clearly say that we started with the dataframe, took a subset of rows, selected the death_rate, used mean to get an average, and then rounded the value.

Note the period . used above in the third line of code is used to reference the object being piped by magrittr and so above it refers to the dataframe that the subset function outputs.

The `magrittr` package is a core package that is used in all the tidyverse packages and helps inform the opinionated style of analysis used in the `tidyverse`.

tibble

The `tibble` package is another `tidyverse` foundational package. A `tibble` is the tidyverse version of the base R dataframe. You will not necessarily use this package directly, but the `tidyverse` functions will return `tibbles` instead of dataframes.

In practice, this will not affect you much and you can treat a `tibble` as a dataframe. However, `tibble` will behave in different ways. For example, if we typed in the word iris in the R Console we would simply get a listing of all the contents in the iris dataframe. However, if you convert the iris dataframe to a `tibble` and examine the output like this:

```
iris %>% as_tibble()
```

You will get tidy printout of data that looks like what is pictured in Figure 6-1.

```
> iris %>% as_tibble()
# A tibble: 150 x 5
   Sepal.Length Sepal.Width Petal.Length Petal.Width Species
          <dbl>       <dbl>        <dbl>       <dbl> <fct>
 1          5.1         3.5          1.4         0.2 setosa
 2          4.9         3            1.4         0.2 setosa
 3          4.7         3.2          1.3         0.2 setosa
 4          4.6         3.1          1.5         0.2 setosa
 5          5           3.6          1.4         0.2 setosa
 6          5.4         3.9          1.7         0.4 setosa
 7          4.6         3.4          1.4         0.3 setosa
 8          5           3.4          1.5         0.2 setosa
 9          4.4         2.9          1.4         0.2 setosa
10          4.9         3.1          1.5         0.1 setosa
# ... with 140 more rows
>
```

Figure 6-1. *Tibble tidy output*

This shows only one screen of data, but more useful information is presented. We can see a sample of the data itself and we get the data types for each column and the dimensions of the dataset.

dplyr

dplyr is a tidyverse package that you use to do data manipulation. This package makes it much easier to filter rows, select variables, and mutate content. dplyr works hand in hand with magrittr to provide a "grammar" for data manipulation.

In dplyr and tidyverse in general, the word grammar refers to the opinionated way of working in R. In the tidyverse grammar, functions are called verbs. Verbs are written in a way to make them very descriptive so that when you use tidyverse grammars the code that you are writing tells the story of your analysis in a very clear way.

For example, you can code the last example where we found the average death rate using dplyr and magrittr like this:

```
expectancy %>%
  filter(race == "All Races") %>%
  summarize(avg = mean(death_rate)) %>%
  mutate(avg = round(avg, 1))
```

In the code above, we pipe the expectancy dataframe to a set of verbs that describe what we are doing. We are filtering, summarizing, and mutating. This describes what we are doing. It also makes it easy to make changes to this pipeline, since we can simply change or add new verbs anywhere in the pipeline.

The code above will output one value that gets returned as a tibble.

```
# A tibble: 1 x 1
    avg
  <dbl>
1 1489.
```

If we want to do the same analysis, but also see what the average death rate is by sex we can simply add a group_by verb to the pipeline.

```
expectancy %>%
  filter(race == "All Races") %>%
  group_by(sex) %>%
  summarize(avg = mean(death_rate)) %>%
  mutate(avg = round(avg, 1))
```

This time we will get three values: one for both sexes, one for males, and one for females.

```
# A tibble: 3 x 2
  sex           avg
  <chr>       <dbl>
```

```
1 Both Sexes 1478.
2 Female     1305.
3 Male       1683.
```

SQL Like Joins

Use the inner_join, left_join and right_join verbs to combine two related datasets together. These dplyr verbs are roughly the equivalent of the similarly named SQL statements INNER JOIN, LEFT JOIN, and RIGHT JOIN.

To illustrate a dplyr join let's imagine that we wanted new labels in our dataframe (excuse me, tibble!) and we also had a tibble that included our better labels. You can create a label tibble like this to follow along:

```
labels <- tribble(
    ~key, ~new_label,
    "Both Sexes",   "All Genders",
    "Female",   "Identifies Female",
    "Male",   "Identifies Male"
  )
```

If you run this code, a new tibble named "labels" will appear under "Data" in your environment window. You can view this in the same way as you viewed dataframes before by simply clicking on the object name in the view.

We can now easily add these labels into our pipeline using the inner_join dplyr verb.

```
expectancy %>%
  filter(race == "All Races") %>%
  group_by(sex) %>%
  summarize(avg = mean(death_rate)) %>%
  inner_join(labels, by = c("sex" = "key")) %>%
```

```
  mutate(avg = round(avg, 1)) %>%
  select(gender = new_label,
         avg_death_rate = avg)
```

In the code above, you can also see that we used a `select dplyr` verb to arrange and rename the labels in a more pleasing way. You can see the results in Figure 6-2.

```
> expectancy %>%
+    filter(race == "All Races") %>%
+    group_by(sex) %>%
+    summarize(avg = mean(death_rate)) %>%
+    inner_join(labels, by = c("sex" = "key")) %>%
+    mutate(avg = round(avg, 1)) %>%
+    select(gender = new_label,
+            avg_death_rate = avg)
# A tibble: 3 x 2
  gender              avg_death_rate
  <chr>                    <dbl>
1 All Genders              1478.
2 Identifies Female        1305.
3 Identifies Male          1683.
```

Figure 6-2. *Gender labels applied to tibble*

When you run the code above, you will see that the results are returned as a `tibble`. Instead of displaying the results in the Console you could also use the assignment operator `<-` to store the results so that you can use them later on.

```
avg_death_rates_by_gender <-
  expectancy %>%
  filter(race == "All Races") %>%
  group_by(sex) %>%
  summarize(avg = mean(death_rate)) %>%
  inner_join(labels, by = c("sex" = "key")) %>%
```

```
mutate(avg = round(avg, 1)) %>%
select(gender = new_label,
       avg_death_rate = avg)
```

stringr

The last `tidyverse` package that we will look at in detail will help you work with strings. Working with strings is tricky because this data type can be more complicated that a simple number. `stringr` includes functions to help detect patterns, split strings and join strings.

You can use stringr functions on their own, but also with `dplyr` verbs like `filter` and `mutate`. For instance, to filter records that include the pattern of characters "All" in the race column you use `str_detect` in the filter verb like this:

```
expectancy %>%
  filter(str_detect(race, "All"))
```

We could use `str_replace` to replace one pattern of characters with another. To change the race column "All Races" entry to "Most Races" you could do this:

```
expectancy <-
  expectancy %>%
  mutate(race = str_replace(race, "All", "Most"))
```

There are many `stringr` functions that will help you work with strings. Many are simply wrappers that work with base R functions. The value of `stringr` is mainly in that it provides a consistent interface. Each function will have the input string as the first argument so these can work with the `magrittr` pipe operator.

Conclusion

There are many more packages in the `tidyverse` collection, but the ones covered above will give you the knowledge to be able to work with any `tidyverse` package. These are also the packages that you are most likely to use on a daily basis and require concepts that extend base R enough that we felt that they deserved a more detailed treatment. But, let's also take a moment to mention so other packages just so you are aware of what else is available to you.

`ggplot2` is a `tidyverse` package that really sets the standard for data visualization in R. This package will get detailed treatment in the chapter on data visualization. `tidyr` helps transform data into tidy `tibble` datasets that are easy to add to your analysis pipeline. `readr`, `readxl,` and `haven` helps you import text data, Excel data and SAS/Stata/SPSS data. `lubridate` helps you work with tricky dates and date-times.

CHAPTER 7

Data Visualization

One of the most appealing elements for RStudio users is the rich library of data visualization tools that are available. In this chapter, we are going to highlight major graphic tools that will inform most of your daily work. Of course, in the R ecosystem we have hundreds of additional libraries and specialized packages that may be used for visualization.

For now, we will focus on the tools that will have the most impact on you immediately starting with ggplot2 (a tidyverse package). From there, we will move on to packages that give you more control of your visualizations at the cost of requiring more work from you to use.

ggplot2

ggplot2 is the tidyverse way of going essential plotting. This replaces and enhances the base R methods of plotting data that we have already seen in earlier chapters. As you may expect from our discussion of dplyr in the last chapter, ggplot2 offers a very opinionated way of doing data visualization. You will use ggplot2 for your daily data exploration and also for inclusion in reports where static graphics are acceptable.

© Matthew Campbell 2019
M. Campbell, *Learn RStudio IDE*, https://doi.org/10.1007/978-1-4842-4511-8_7

Let's take a look at how to use ggplot2. Firstly, as with the other tidyverse packages we need to use the library function to load the tidyverse packages library(tidyverse).

```
library(tidyverse)
```

```
source("process_data.R")
```

After we loaded tidyverse, in the second line of code, we used the source command to rerun all the code that we saved in the "process_data.R" file. This ensures that we are both on the same page when it comes to the structure of the expectancy dataset.

To use ggplot2, we can pipe a dataset to the ggplot function. This function requires us to specify the variables that we want to plot.

```
expectancy %>%
   filter(race == "All Races",
          sex == "Both Sexes") %>%
   ggplot(aes(year, life_expectancy))
```

When you run this code, you will see a blank canvas appear with the X and Y axis already defined as pictured in Figure 7-1.

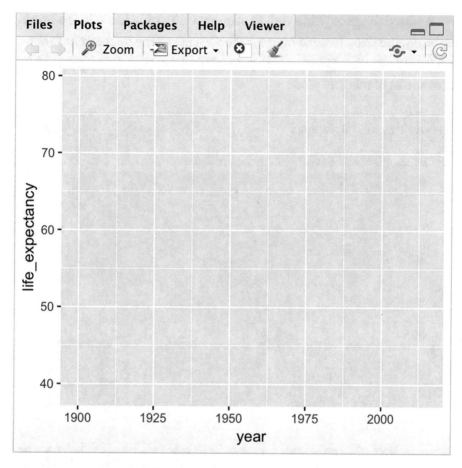

Figure 7-1. *Blank ggplot canvas*

Figure 7-1 shows a canvas with the year on the X axis and the death rate on Y axis. Now that we have defined the scope of the plot, we can start to add visual elements to show the data. You do that by piping the ggplot function to a set of "verbs" that describe the data story.

```
expectancy %>%
  filter(race == "All Races",
         sex == "Both Sexes") %>%
  ggplot(aes(year, life_expectancy)) +
  geom_line()
```

We used the geom_line verb to state that we would like to use a line to represent the data. Instead of using the magrittr pipe operator %>% that you would expect, ggplot2 uses a plus symbol + to pipe. See Figure 7-2 for an example of this line plot.

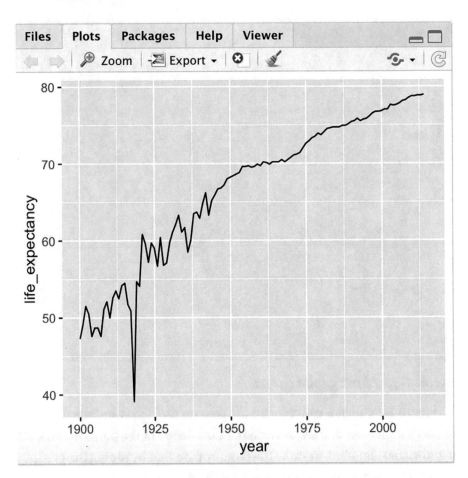

Figure 7-2. *Death rate by year line plot*

The plot in Figure 7-2 shows an interesting trend as well as a possible outlier that may need to be explained later on. You can continue to layer on visualizations by piping them using the pipe operator +. If you wanted to add points to the line you would use the geom_point verb.

```
expectancy %>%
  filter(race == "All Races",
         sex == "Both Sexes") %>%
  ggplot(aes(year, life_expectancy)) +
  geom_line() +
  geom_point()
```

This would produce the plot show in Figure 7-3.

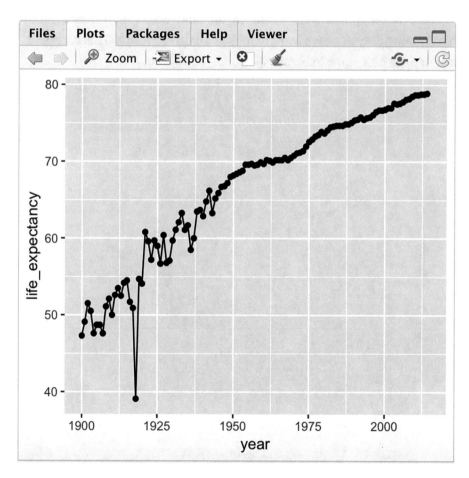

Figure 7-3. *Adding point markers to line plot*

What if we wanted to see this trend broken down by sex? To look at subgroups like that in ggplot line plot you simply add a new aesthetic parameter to the geom_line verb that you want to categorize.

```
expectancy %>%
  filter(race == "All Races") %>%
  ggplot(aes(year, life_expectancy)) +
  geom_line(aes(color = sex))
```

Note that we are no longer filtering on "Both Sexes" in the first part of this visualization since we want to look at all three groups. This code will produce the plot show in Figure 7-4.

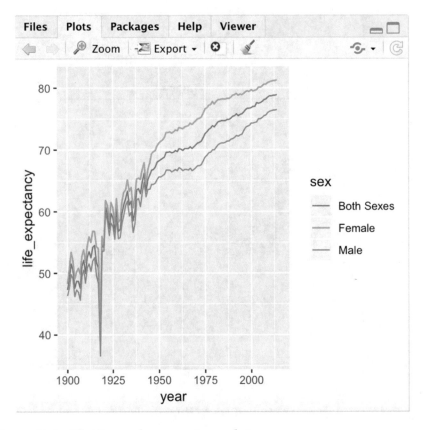

Figure 7-4. *Plotting subgroups in ggplot*

This plot shows us the relationship between the two sexes that we have data on. The lines appear in color because we specified the color aesthetic in our geom_line.

There is much more to ggplot that we have space to cover in this chapter. You get plenty of options to display different types of data along with label and title options. This package is essentially the visualization workhorse for RStudio. Take a look at the official website ggplot2 to get more information on how to use this package: https://ggplot2. tidyverse.org.

htmlwidgets

While ggplot2 provides most of what you will need with essential plotting, there are times when you may want to add interactivity or very specialized graphics to your analysis. htmlwidgets are a collection of packages that you can use to add this next level of visualization to your R analysis.

htmlwidgets gives you the ability to incorporate open source Javascript widgets that were originally developed for websites. Go to the http:// htmlwidgets.org website to get a complete list of the widgets available to you. Note that many of these widgets are very specific to the type of problem that you are trying to visualize. You will find widgets for maps, time-series, network graphs and some D3 graphics.

We can use one htmlwidget, plotly, to turn the plot we created in the last section into a web ready interactive widget. Plotly is interesting because it works together with gglot2 and really takes ggplo2t to the next level. Here is all you need to do to add interactivity to a ggplot2 plot with plotly.

```
library(plotly)

i_plot <-
  expectancy %>%
  filter(race == "All Races") %>%
```

```
ggplot(aes(year, life_expectancy)) +
geom_line(aes(color = sex))
```

```
ggplotly(i_plot)
```

This code is very similar to what we did with ggplot2. However, this time we are assigning the results to an object and using that object as a parameter to ggplotly to create the interactive plot you see in Figure 7-5.

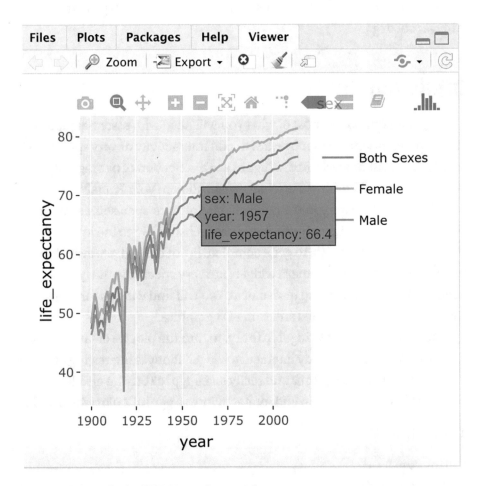

Figure 7-5. *Plotly HTML widget with popup*

This plot now includes detail popups and a toolbar with navigation figures at the top. In a later chapter, when we learn how to create cohesive reports and dashboards, we will be able to include widgets like this to create a rich interactive experience for our users.

Before we move on, let's highlight the DT package Datatables component. Datatables is a remarkably useful interactive html widget that will create a searchable HTML presentation of your datasets. For instance, if we wanted to provide a searchable version of our dataset directly to our users we could simply do this:

```
library(DT)
datatable(expectancy,
          options = list(pageLength = 10))
```

This will create the widget shown in Figure 7-6, which is a searchable interactive table filled with our data.

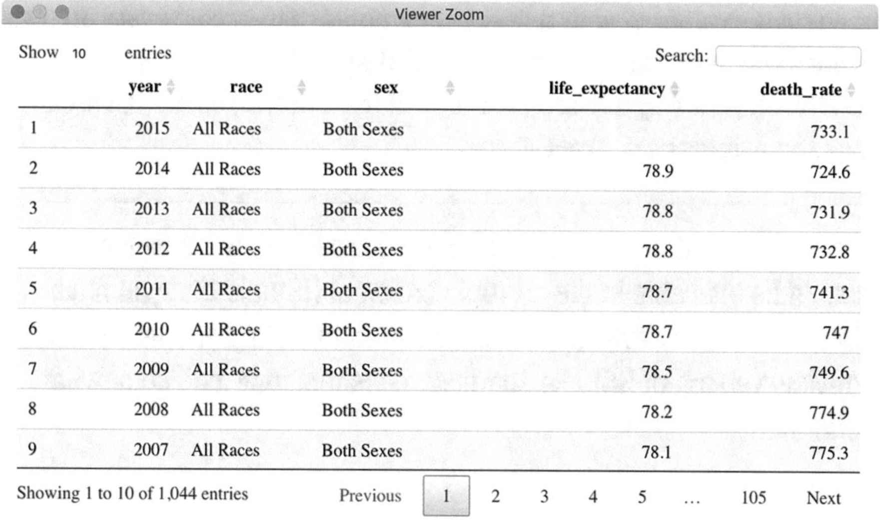

	year	race	sex	life_expectancy	death_rate
1	2015	All Races	Both Sexes		733.1
2	2014	All Races	Both Sexes	78.9	724.6
3	2013	All Races	Both Sexes	78.8	731.9
4	2012	All Races	Both Sexes	78.8	732.8
5	2011	All Races	Both Sexes	78.7	741.3
6	2010	All Races	Both Sexes	78.7	747
7	2009	All Races	Both Sexes	78.5	749.6
8	2008	All Races	Both Sexes	78.2	774.9
9	2007	All Races	Both Sexes	78.1	775.3

Showing 1 to 10 of 1,044 entries   Previous 1 2 3 4 5 ... 105 Next

Figure 7-6. Data table widget

r2d3

The packages discussed in the first two sections of this chapter all provide rich objects to visualize most types of data. However, at the end of the day our graphics will look like the graphics that were designed by others and are available to all R developers.

There are some use cases where we want to have absolute control over every aspect of a visualization. Either a widget doesn't exist that would help us or we need to something which is highly customized. This is where D3 can help us and r2d3 is a package that we can use to incorporate D3 into our RStudio project.

Like htmlwidgets, D3 is a web based Javascript solution. However, using D3 is more involved that what we would do with htmlwidgets. D3 is essentially a programming language.

The core idea in D3 is that you can bind datasets with graphic primitives (things like lines and circles). These graphics will grow or shrink based on the values in your dataset. D3 graphics can become very involved because you have a lot of control on how they look.

While it goes way beyond the scope of this book to teach D3, we can show you a quick hack to get started.

Note   The features discussed here require RStudio 1.2 which should be available in the current version of RStudio that you have downloaded. However, the version used in this chapter was from a Preview version of RStudio so the screenshots may look somewhat different.

To use D3, simply go to the RStudio menu and then choose "File", "New File", and then "D3 Script". A file will appear with D3 code.

```
// !preview r2d3 data=c(0.3, 0.6, 0.8, 0.95, 0.40)
//
// r2d3: https://rstudio.github.io/r2d3
//

var barHeight = Math.ceil(height / data.length);

svg.selectAll('rect')
 .data(data)
 .enter().append('rect')
 .attr('width', function(d) { return d * width; })
 .attr('height', barHeight)
 .attr('y', function(d, i) { return i * barHeight; })
 .attr('fill', 'steelblue');
```

The code in this file is Javascript and you can adjust the visualization by modifying this code. Save this file and click the "Preview" button to see what the D3 visualization looks like. It doesn't look like much compared to the pre-packaged solutions that we highlighted earlier, but that is ok since D3 really offers more of a canvas for you to draw on and not a set of plot widgets.

The preview simply uses the code found in the comments in the file, but if you rather provide your own data from a R script file you can do this using the r2d3 library.

```
library(r2d3)

r2d3(data = c(.1,.2,.3,.4,.3,.2,.1),
     script = "d3_plot.js")
```

This code will produce the plot in Figure 7-7.

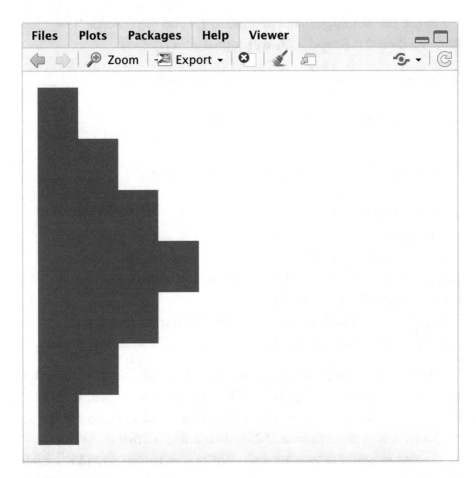

Figure 7-7. *Basic D3 plot*

Clearly, this is a very basic beginning visualization. Check out the r2d3 website for a whole set of stunning visualizations that you can do with D3 https://rstudio.github.io/r2d3/articles/gallery.html. These examples also contain the code that you need to recreate them so that you can use this as a starting point for your own D3 custom visualizations.

Conclusion

In addition to the base R `plot` function, you have many options for data visualization in R. ggplot provides a very robust and opinionated way of doing what most analysts need most of the time. Packages like `r2d3` and `htmlwidgets` provide support for more specialized use cases. There are even more specialized visualization options available and all of these features may be used in the dashboards and reporting options that we will discuss in later chapters.

CHAPTER 8

R Markdown

You use R Markdown to create reports that incorporate the features discussed in the previous chapters. R Markdown leverages freely available open source technology like HTML, CSS, and Markdown to create rich reports in a variety of formats.

R Markdown is based on Markdown, which is a short-hand method of defining a document in a human readable way. The idea is that you can take quick notes in a structured way. These notes are readable on their own, but they can also be translated into polished presentation formats such as web pages, presentations, and PDF files.

RStudio extends Markdown by adding in code and graphics from the R language into Markdown documents. This type of Markdown is referred to as "R Markdown". Let's use our life expectancy analysis to create a shareable report using R Markdown.

R Markdown Documents

R Markdown is completely supported in RStudio. Not only can you write your notes in Markdown, but code that you include can be executed in blocks and displayed right on your R Markdown document in real time. To create a new R Markdown document, go to the RStudio menu bar and click "File", "New File", and "R Markdown" as shown in Figure 8-1.

© Matthew Campbell 2019
M. Campbell, *Learn RStudio IDE*, https://doi.org/10.1007/978-1-4842-4511-8_8

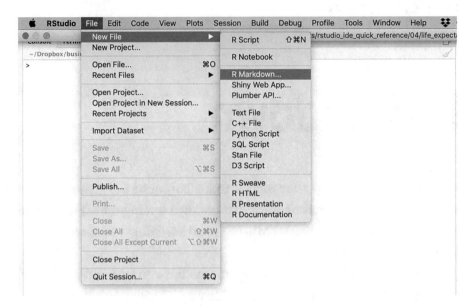

Figure 8-1. *Adding a new R Markdown document*

In the dialog box that appears as pictured in Figure 8-2, you will be presented with options that will determine the type of document that gets generated.

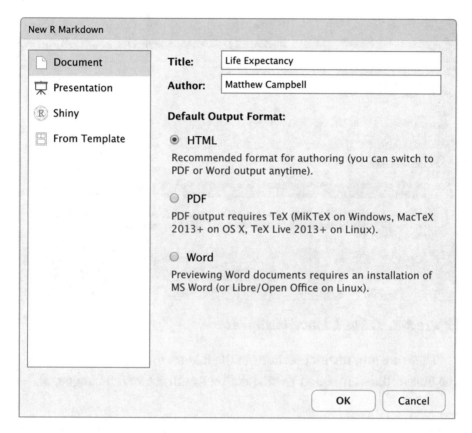

Figure 8-2. *R Markdown options*

Here is where you specify whether you want an HTML or PDF document. There are some other less used format types as well that you can experiment with, but generally people use PDF or HTML reports.

RStudio will automatically generate a R Markdown document with template text that resembles Figure 8-3.

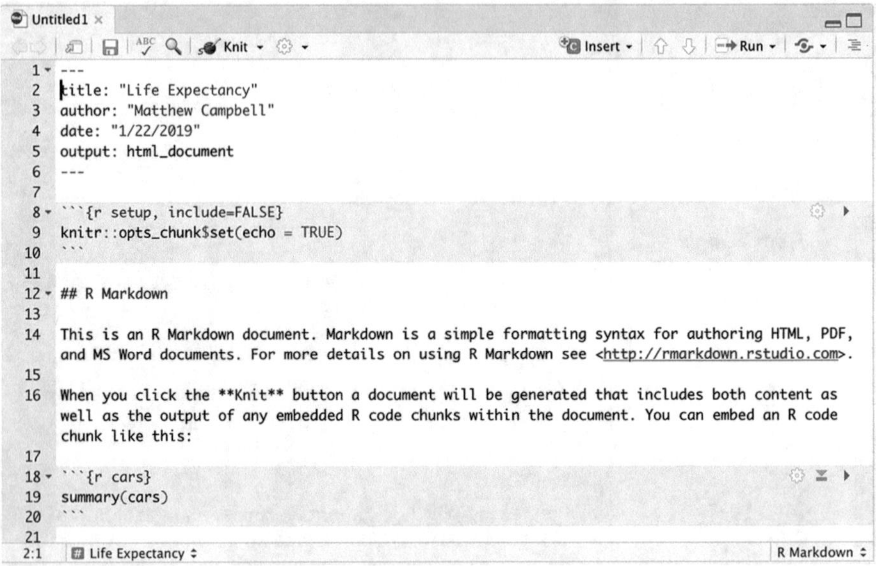
Untitled1
Knit
Insert
Run
---
title: "Life Expectancy"
author: "Matthew Campbell"
date: "1/22/2019"
output: html_document

```{r setup, include=FALSE}
knitr::opts_chunk$set(echo = TRUE)
```
R Markdown
This is an R Markdown document. Markdown is a simple formatting syntax for authoring HTML, PDF, and MS Word documents. For more details on using R Markdown see <http://rmarkdown.rstudio.com>.
When you click the **Knit** button a document will be generated that includes both content as well as the output of any embedded R code chunks within the document. You can embed an R code chunk like this:
```{r cars}
summary(cars)
```
2:1
Life Expectancy
R Markdown

Figure 8-3. *R Markdown document*

There are four distinct sections in the R Markdown document above. The first six lines are called YAML (aka Yet Another Markup Language).

```
---
title: "Life Expectancy"
author: "Matthew Campbell"
date: "1/22/2019"
output: html_document
---
```

This YAML markup specifies things like the type of output format and author name, essentially what you filled out in the dialog box. The next three lines highlighted in gray are used to set options required for knitr, a packaged used to render R Markdown output.

````
```{r setup, include=FALSE}
knitr::opts_chunk$set(echo = TRUE)
```
````

The first line here with the three backticks and curly brackets deserves some attention. This is the notation that you use to escape the Markdown code so that you can include R code in your document. The format is roughly three backticks, an opening curly bracket, the letter r followed by a few options and an ending curly bracket. In the code above, setup is simply a label while include=FALSE means that the code will be executed but not displayed in the R Markdown output.

The next few lines of code are basic Markdown.

```
## R Markdown
```

```
This is an R Markdown document. Markdown is a simple formatting
syntax for authoring HTML, PDF, and MS Word documents. For more
details on using R Markdown see <http://rmarkdown.rstudio.com>.
```

This is where you write up your findings, procedures, and descriptions of your data analysis. The two number signs at the beginning mean "Heading 2" and formatting will reflect this in the output document. You can use all the normal Markdown features. The creator of Markdown has documented the formatting features available to you here: https:// daringfireball.net/projects/markdown.

The remaining code is another R chunk; this time the code creates a plot and is embedded in the R Markdown output.

```
```{r cars}
summary(cars)
```
```

This will output a summary of the built-in cars dataset. We will see an example of what this looks like in the example covered next.

R Markdown Example

Let's illustrate R Markdown by adding some of the work we have done throughout this book into a report. Along the way, we will source R script files and imbed graphics into the report. Start by saving the new R Markdown document that we created with the name "life_expectancy.Rmd".

We can start by replacing the boilerplate introduction provided in the template with something that describes our work.

```
## Introduction

[Life expectancy data](https://catalog.data.gov/dataset/age-
adjusted-death-rates-and-life-expectancy-at-birth-all-races-
both-sexes-united-sta-1900) was downloaded from data.gov. This
dataset contains death rates and life-expectancy statistics
from 1900 to 2015. Exploratory analysis was conducted on this
dataset.
```

Now let's see what the R Markdown output file looks like. Select the menu bar at the top of the R Markdown document and then click the "Knit" dropdown and then select "Knit for HTML" as pictured in Figure 8-4.

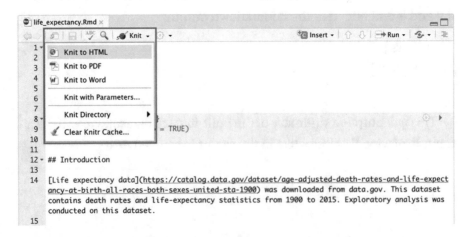

Figure 8-4. *Create HTML output file*

This will take your R and Markdown code and use the R package `knitr` to create an HTML report as shown in Figure 8-5.

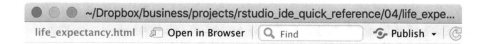

Life Expectancy

Matthew Campbell

1/22/2019

Introduction

Life expectancy data was downloaded from data.gov. This dataset contains death rates and life-expectancy statistics from 1900 to 2015. Exploratory analysis was conducted on this dataset.

```
summary(cars)
```

```
##      speed           dist
##  Min.   : 4.0   Min.   :  2.00
##  1st Qu.:12.0   1st Qu.: 26.00
##  Median :15.0   Median : 36.00
##  Mean   :15.4   Mean   : 42.98
##  3rd Qu.:19.0   3rd Qu.: 56.00
##  Max.   :25.0   Max.   :120.00
```

Figure 8-5. *R Markdown HTML output*

We now have the beginnings of a nicely formatted report. Of course, we still have the cars dataset filler showing, but we will next that in the next step.

R Markdown Reproducibility

When we say our analysis is reproducible, we mean that we should be able to rerun a report and get exactly the same result as we had in the past. Our analysis should not depend on any ad-hoc commands we typed into the command line. In order to make sure our analysis is reproducible, lets run the source command and load the file we used to import our expectancy dataset. Replace the r code block with this and include a summary of our dataset.

````
```{r get_data}

source("process_data.R")

summary(expectancy)

```
````

We did a few things in the code block above. We changed the label to "get_data", we used source to execute all the code saved in "process_data.R" and we are showing a summary of the expectancy dataset. When we knit the Markdown file, the summary of the expectancy dataset will appear instead of the cars dataset.

More importantly, the other code in the R Markdown document will be able to reference in the expectancy dataset in the same way we did earlier. All the data processing rules we followed before will be executed each time we knit a new output document.

You might want to offer your users a glimpse into the dataset like we did in the Data Visualization chapter. You can simply copy your code from the last chapter and wrap it in a code block like this:

```
## Expectancy Dataset Sample

Below you will find a widget that you may use to few the
contents of the data expectancy dataset. This is the dataset
that will be explored later on in this analysis.
```

````
```{r}

library(DT)

datatable(expectancy,
          options = list(pageLength = 5))

```
````

Here we mix Markdown commentary along with a block of r code to produce the display that we can see in Figure 8-6.

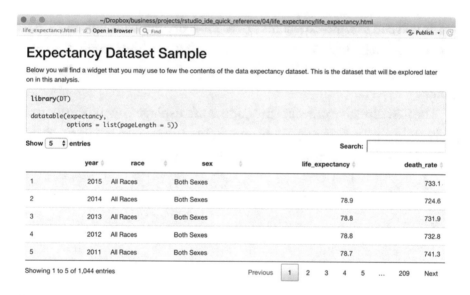

Figure 8-6. *HTML data table widget in R Markdown document*

Finally, let's add the plot of life expectancy rates over years. This time, we may want to hide the r code block from our users so we will include these two directives echo=FALSE and message=FALSE.

Life Expectancy Plot

The plot below shows life expectancy by year for all sexes.

````
```{r echo=FALSE, message=FALSE}
````
````

```
library(plotly)

i_plot <-
  expectancy %>%
  filter(race == "All Races") %>%
  ggplot(aes(year, life_expectancy)) +
  geom_line(aes(color = sex))

ggplotly(i_plot)
```
```

As you can see, life expectancy rises for everyone from 1900 to 2015. Women have a higher life expectancy than men and both groups follow the same upward trend while maintaining a difference between the two groups.

The first directive prevents the code from displaying while the second prevents any messages that might appear because we are loading a library in the document. The result is pictured in Figure 8-7.
```

Life Expectancy Plot

The plot below shows life expectancy by year for all sexes.

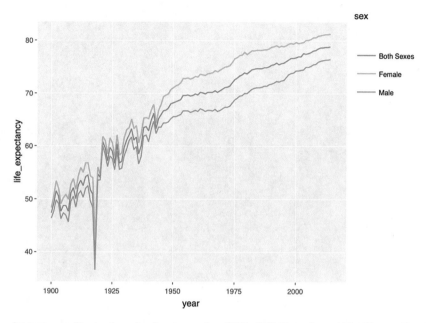

As you can see, life expectancy rises for everyone from 1900 to 2015. Women have a higher life expectancy than men and both groups follow the same upward trend while maintaining a difference between the two groups.

Figure 8-7. *Plotly widget in R Markdown document*

Conclusion

R Markdown provides a powerful report generation system. This chapter has only scratched the surface to what is available to you as a report writer. This will be your reporting workhorse as you develop analytics that may be shared throughout your organization.

CHAPTER 9

Shiny R Dashboards

R Markdown and `htmlwidgets` will give you the ability to create rich and even interactive experiences with popovers. However, some use cases require more than what these types of static tools can provide. For instance, you may need to ask users to supply parameters that will be used in an analysis, connect to a backend database, or execute new R scripts as the user explores your analysis.

You can use Shiny in RStudio to create these types of data dashboards that have a rich functionality that is completely integrated with your R system. Shiny is a package used to create interactive web apps in R. These apps can be used in an RStudio project, added to an R Markdown report, or published online as a web app. Shiny provides an easy to use framework to develop these apps, and you can extend Shiny even further with your own CSS and JavaScript code.

New Shiny Apps

It is very easy to add a Shiny app to your RStudio project. Simply go to the RStudio menu bar and then choose "File", "New File", and then "Shiny Web App...". A dialog box like the one pictured in Figure 9-1 will appear.

© Matthew Campbell 2019
M. Campbell, *Learn RStudio IDE*, https://doi.org/10.1007/978-1-4842-4511-8_9

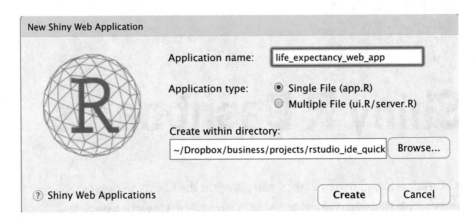

Figure 9-1. *New Shiny app*

You can name your app and choose if you want the code in one or two files. Click the "Create" button to build the web app.

RStudio will create a new folder in your project with the name that you provided in the dialog. If you choose "Single File", a new file named "app.R" will appear and automatically open in your editor view as pictured in Figure 9-2.

```
 1  #
 2  # This is a Shiny web application. You can run the application by clicking
 3  # the 'Run App' button above.
 4  #
 5  # Find out more about building applications with Shiny here:
 6  #
 7  #    http://shiny.rstudio.com/
 8  #
 9
10  library(shiny)
11
12  # Define UI for application that draws a histogram
13  ui <- fluidPage(
14
15    # Application title
16    titlePanel("Old Faithful Geyser Data"),
17
18      # Sidebar with a slider input for number of bins
19      sidebarLayout(
20          sidebarPanel(
21              sliderInput("bins",
22                          "Number of bins:",
23                          min = 1,
24                          max = 50,
```

Figure 9-2. *app.R contents*

The code that appears is a template that RStudio creates for you as a starting point. This is a working Shiny App that uses the built-in R datasets just as an example. Click the "Run App" button shown in Figure 9-2 to run the web app. The app will start to run in a local web server that is managed by R as you can see pictured in Figure 9-3.

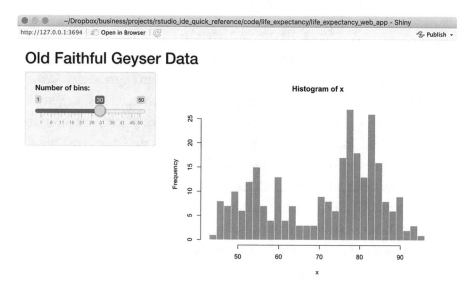

Figure 9-3. *Shiny App running locally*

This histogram is very interactive, when you adjust the slider control on the left R code is executed to recreate the histogram on the right. This is a step beyond simply including popovers that are pre-baked into a static report.

Along the top of Figure 9-3, you will see an http address and an "Open in Browser" button. This app is running in a web server, so you can view the app using your regular browser by either clicking the "Open in Browser" button or typing in the http address into your browser's address bar.

You may also notice that there is a "Publish" button at the right. Shiny apps are published to a web site so that they can be made available for anyone to use. These apps can be public either with the service that

RStudio provides or you can manage a private server from your business. To get a free account to publish your apps, head to `https://www.shinyapps.io` to sign up.

Understanding Shiny

Shiny can be used to create dashboards that use R scripts and are accessible on the web fairly easily. However, like most R packages you can go much deeper into the system to build truly custom experiences. We will discuss the basics here to give you a good starting point with Shiny, but to really master this package head over to `https://shiny.rstudio.com/` to see tutorials and documentation that shows what is possible with Shiny.

Shiny code is separated into two areas of distinct responsibility: the user interface and the server. This is also called the "front end" and the "back end". The idea is that the visual user interface elements like sliders and buttons belong to front end part of the app, while the number crunching and data analysis belong to the back end part of the app.

In the code we generated, there is only one file that contains both the user interface and the server. You will also see these two components separated out in their own file depending on what options you choose or who originally coded the app.

User Interface

The user interface is contained in the `ui` object coded in the "app.R" file that was created for us. The code that creates the `ui` object is located toward the top of the app.R file:

```
ui <- fluidPage(

    # Application title
    titlePanel("Old Faithful Geyser Data"),
```

```
# Sidebar with a slider input for number of bins
sidebarLayout(
    sidebarPanel(
        sliderInput("bins",
                    "Number of bins:",
                    min = 1,
                    max = 50,
                    value = 30)
    ),

    # Show a plot of the generated distribution
    mainPanel(
        plotOutput("distPlot")
    )
  )
)
```

This entire block of code is only one function fluidPage that is
called with a comma-separated list of two parameters titlePanel and
sidebarLayout. The results of this function are assigned to the ui object.
This ui object will be used toward the end of this file as a parameter to the
function that starts the Shiny app.

Take a moment to slow down and really look at the code above. This
is probably not how you are used to writing code. The words used for the
function names suggest what is happening. If you think about it, you might
expect that we will be looking at a page that contains a title followed by
something more complicated (ignore that for now). You can clearly see
where we are getting the title "Old Faithful Geyser Data" that we first saw in
Figure 9-3.

The essential process to building the Shiny front end is here. If you
wanted to add something to the app, you would include a new function
call in the comma-separated list. For instance, if I wanted to add a HTML

paragraph to explain what Old Faithful was, I could add the p function to the space right after the `titlePanel`.

```
# Define UI for application that draws a histogram
ui <- fluidPage(

    # Application title
    titlePanel("Old Faithful Geyser Data"),

    p("Old Faithful is a cone geyser located in Yellowstone
National Park in Wyoming, United States. It was named in 1870
during the Washburn-Langford-Doane Expedition. Wikipedia
2019."),
```

The result would be the web app that you can see in Figure 9-4. It is essentially the same app, but now it includes some text context.

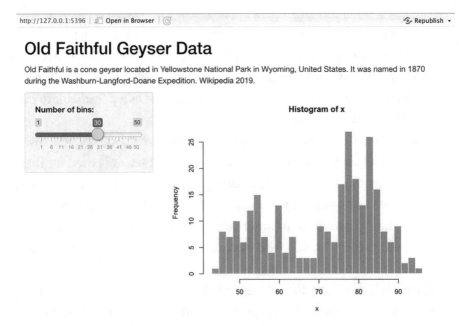

Figure 9-4. *Adding a paragraph to a Shiny app*

As you build out your app, you will continue to add pieces into the page. Some functions are simple HTML components like the p function, and some are specific to Shiny. Often, you will find objects nested within other objects in order to define a layout for the app.

You can see this in sidebarLayout, an object in the overall fluidPage that is in turn composed of two other objects that appear as parameters in a comma-separated list.

```
# Sidebar with a slider input for number of bins
sidebarLayout(
    sidebarPanel(
        sliderInput("bins",
                    "Number of bins:",
                    min = 1,
                    max = 50,
                    value = 30)
    ),

    # Show a plot of the generated distribution
    mainPanel(
        plotOutput("distPlot")
    )
)
```

The two components above are sidebarPanel and mainPanel. sidebarPanel contains yet another nested object, sliderInput, that defines the slider that the user will use to control the visualization. The very first parameter value bins is called an inputID and is used to identify the component on the server backend of the web app.

mainPanel contains plotOutput which is the object that is presenting the graphic. In plotOutput the parameter is highlighted because this is going to be used as the inputID so that the server component can identify this object.

To conclude, the `ui` object now contains objects that display a title, a paragraph, a slider control, a plot, and other objects to enforce a layout. Two of the objects, the slider and the plot, also have `inputID` parameters so that the server will be able to retrieve values from the slider and know where to present the graphic.

Server

The server component takes care of the heavy lifting on the backend. In this example, server will gather the parameters that the user defined on the user interface to run R code that will do the analysis and create the plot. `inputID` is used to identify both the input and the output components. Here is the server code with the input and outputs highlighted.

```
# Define server logic required to draw a histogram

server <- function(input, output) {

    output$distPlot <- renderPlot({
        # generate bins based on input$bins from ui.R
        x    <- faithful[, 2]
        bins <- seq(min(x), max(x),
                    length.out = input$bins + 1)

        # draw the histogram with the specified
        # number of bins
        hist(x, breaks = bins,
             col = 'darkgray',
             border = 'white')
    })
}
```

In the code above, we are assigning an anonymous function to the server object. These functions are a little bit different than what we did in the user interface above. In the user interface, we simply assigned the results of a function call to the ui object while here we are assigning a set of instructions in the form of an anonymous function to the server object. These instructions will be used to respond to the activity in the user interface.

The code in bold are the hooks where the server connects to the user interface. Everything is referenced in the anonymous function's input and output parameters. input contains references to every input component in the user interface, while output contains references to every output component in the user interface.

You reference an input component using the input object followed by a dollar sign and the inputID of the input object input$bins. If you remember, "bins" was the identifier given to the slider object. input$bins will return the current value that the user has chosen.

The output object here is getting all the code that is generated by the renderPlot function above. You can see that we reference this output component in a similar way using the function's parameters output$distPlot. You can see the mechanics of how the user input is gathered and the plot displayed in the lines of code contained in the renderPlot function.

App

There is one more thing that has to happen before Shiny can create the web app. We need to associate the ui and server components and then launch them with this code shinyApp(ui = ui, server = server).

This is included as boilerplate code and you may not have to touch this code at all unless you change the name of the user interface and server objects. But, it's worth knowing where this code is located and that this code launches the app instance.

Customizing Shiny Apps

Let's change this boilerplate code into something that works for our analysis. For now, let's just make some minor adjustments to the user interface and let's display our data instead of the built it `faithful` dataset.

The first thing we need to do is copy our dataset into the Shiny folder so we can deploy this with all the assets in one place in the future. Save the dataset as an rdata file in the Shiny app folder using this code on the R command line:

```
save(expectancy, file = "life_expectancy_web_app/expectancy.
rdata")
```

Now we can load this into our Shiny session when the app launches. Put code at the top of the Shiny app.R file to load the dataset.

```
load("expectancy.rdata")
```

```
expectancy <-
  expectancy %>%
  filter(!is.na(life_expectancy))
```

This loads the dataset and then removes any missing values that will interfere with the histogram. Now, we can look at our dataset instead of faithful but changing one line of code in the `server` component.

```
server <- function(input, output) {

    output$distPlot <- renderPlot({
        # generate bins based on
        # input$bins from ui.R
        x <- expectancy$life_expectancy

        bins <- seq(min(x), max(x),
                    length.out = input$bins + 1)
```

```
        # draw the histogram with the
        # specified number of bins
        hist(x, breaks = bins,
            col = 'darkgray',
            border = 'white')
    })
}
```

Before we test this out, let's make changes to the title description and add our own colors into the plot. You can probably guess where we will make these changes. In the code below all the changes are highlighted in bold.

```
library(shiny)
library(tidyverse)

load("expectancy.rdata")

expectancy <-
  expectancy %>%
  filter(!is.na(life_expectancy))

# Define UI for application that draws a histogram
ui <- fluidPage(

    # Application title
    titlePanel("Life Expectancy Data"),

    p("Life expectancy data was downloaded from
        data.gov. This dataset contains death rates
        and life-expectancy statistics
        from 1900 to 2015. Exploratory analysis was
        conducted on this dataset. In the
        visualization below, you can see all life
        expectancy  rates in the dataset and adjust
```

**the number of bins used to present the
frequency.**"),

```
    # Sidebar with a slider input for number of bins
    sidebarLayout(
        sidebarPanel(
            sliderInput("bins",
                        "Number of bins:",
                        min = 1,
                        max = 50,
                        value = 30)
        ),

        # Show a plot of the generated distribution
        mainPanel(
            plotOutput("distPlot")
        )
    )
)

# Define server logic required to draw
# a histogram

server <- function(input, output) {

    output$distPlot <- renderPlot({
        # generate bins based on
        # input$bins from ui.R
        life_expectancy <- expectancy$life_expectancy

        bins <- seq(min(life_expectancy),
                    max(life_expectancy),
                    length.out = input$bins + 1)
```

```
        # draw the histogram with the
        # specified number of bins
        hist(life_expectancy,
              breaks = bins,
              col = 'lightblue',
              border = 'white')
    })
}

# Run the application
shinyApp(ui = ui, server = server)
```

Now when you run this code, your app will resemble the one that appears in Figure 9-5.

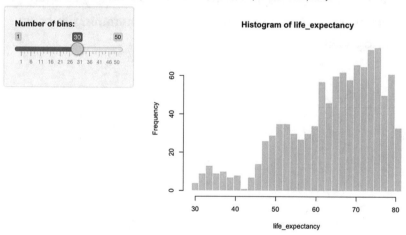

Figure 9-5. *Life Expectancy Shiny App*

The context now is relevant, and we also adjusted the color of the bars in the histogram. You can see a published version of this on the author's shiny.io page by navigating to this URL: `https://mattjcamp.shinyapps.io/life_expectancy_web_app`.

Conclusion

In this chapter, we introduced Shiny and really just scratched the surface of what we can accomplish with this tool. Shiny creates a pathway to deploy data-rich apps directly to the web all without requiring a specific background in web development.

That being said, Shiny is a very rich package and even without dipping into JavaScript you can find yourself spending a good deal of time working on data web apps. Take some time to learn the basics and then follow the pattern in this chapter with example code from `https://shiny.rstudio.com`. Look for examples that you find interesting that would work with data that you are familiar with and adjust the code to make your own visualizations. As you do this and learn the patterns, deploying Shiny apps will become second nature to you as statistical programming and data analysis is to you now.

CHAPTER 10

Custom R Packages

R owes its popularity in the data field largely due to its extensive library of third-party packages that add significant functionality to the core R language. We have already seen how using the `tidyverse` package can transform the experience of using R. Many data problems have already been solved and R programmers make these solutions available for anyone to use. In fact, many data projects are started with a search for R packages that already exist that solve the problem at hand.

In this chapter, we are going to talk about how we can create our own R packages either for own teams or for distribution to the wide world.

Custom R Packages Use Cases

Why should we create our own packages? The obvious reason to create a package is to share our work with the world. Problems that have been solved in R and can be coded in a way that is generally useful are good candidates for packages.

You may also want to create packages that only really apply to your own team or you own projects. The reason is that coding in a package format encourages code reuse and process efficiency. You may also use packages to bundle data and code together for your colleagues, especially if you are seeing similar situations cropping up. Finally, R packages come with built-in documentation tools that provide a way to communicate complex material in your work.

© Matthew Campbell 2019
M. Campbell, *Learn RStudio IDE*, https://doi.org/10.1007/978-1-4842-4511-8_10

Not all code belongs in a package, so you will have to apply some judgement when it comes to turning an ad-hoc analysis into a package. Generally, it is more difficult to iterate and test code when an R package format, so it is usually best to wait until you are in maintenance mode to package up a set of functions into a package.

In this chapter, we will create a package based on the work we have been doing in the previous few chapters. For this work, we will use RStudio to create a completely new project and then selectively bring in the features for the package.

Create New Custom R Package

RStudio has built in support for creating packages. Go to the RStudio menu bar and then choose "File", "New Package", "New Directory", and then "R Package". A dialog box like Figure 10-1 will appear.

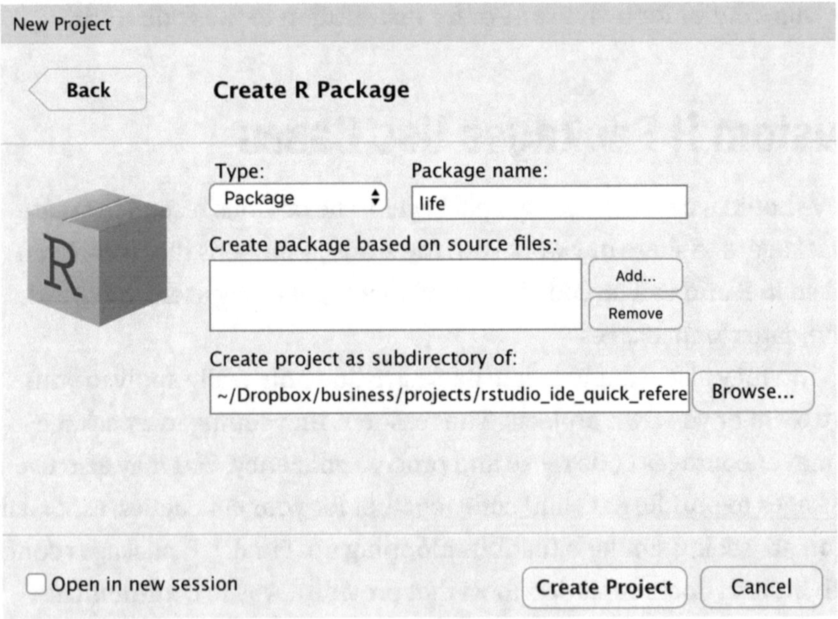

Figure 10-1. *R Package creation dialog box*

You can add a name for your package and specify a directory to create the package in. Click the "Create Project" button to create the R package project.

You will get a set of template files in your folder that you can follow as a guide. The key files are "hello.R" in the "R" folder, "DESCRIPTION" and "hello.Rd" in the "man" folder. Let's talk about each of these below.

hello.R

All of the code that you want to publish should be located in the "R" folder just as "hello.R". This file will be opened when you create the R package project and you can see these contents.

```
# Hello, world!
#
# This is an example function named 'hello'
# which prints 'Hello, world!'.
#
# You can learn more about package authoring with
# RStudio at:
#
#   http://r-pkgs.had.co.nz/
#
# Some useful keyboard shortcuts for package
# authoring:
#
#   Install Package:         'Cmd + Shift + B'
#   Check Package:           'Cmd + Shift + E'
#   Test Package:            'Cmd + Shift + T'

hello <- function() {
  print("Hello, world!")
}
```

This is very simple, but good enough to demonstrate how the R package build process works. You can put the functions you want to publish here to make them available. It is a good practice to include one file per function in the "R" folder.

DESCRIPTION

The "DESCRIPTION" file provides the meta-data for the package. This is where you document the title, description, version, and your contact information. Simply open the file by clicking on it and replace any fields with information about your R package.

```
Package: life
Type: Package
Title: Shows Life Expectancy Trends Since 1900
Version: 0.1.0
Author: Matt
Maintainer: Matt <matt@somewhere.net>
Description: Includes datasets and helper functions to help
             present information on life expectancy rates.
License: None
Encoding: UTF-8
LazyData: true
RoxygenNote: 6.1.1
```

You don't need to change this to follow along with the demo, but when you are ready to publish you want to make sure that these fields contain relevant information.

hello.Rd

"hello.Rd" is the R documentation file. This file is automatically generated for you based on documentation that you include with your function. We will talk more about documenting code below.

The information here will appear in the package documentation that you read in the "Help" pane. The title will also appear as part of the code completion in RStudio so that when you are typing out a line of code your content in appear right in the editor screen. This is a really handy way of providing context to other analysts on your team right when they may need it. Let's build the R package so we can see how this all works.

Build R Package

When you are working with an R package project, you will have an additional tab in your top right view called "Build". This tab gives you some options around building R packages. Go to this tab and then click "More" and then "Clean and Rebuild" as pictured in Figure 10-2.

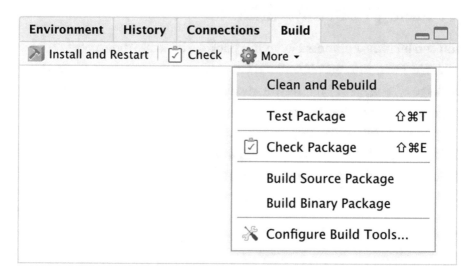

Figure 10-2. *Build view*

This will build and package up your code. If there are any required libraries that are needed but are not installed RStudio will prompt you to install them now. Follow the dialog wizard to install any required packages.

Now that you have built and installed your package, you may use the function like this:

```
library(life)
```

```
hello()
```

This simple code will print "Hello, world!" to your Console. If wanted to change the behavior of this function, you can simply change the code in the "hello.R" file and then rebuild the package. Go ahead and try to change this function and see the results.

You may also use these functions in the same way from your other RStudio projects. This is how R packages encourage code reuse. Let's go ahead and add some features from our life expectancy analysis to show off R packages a bit more.

R Package Documentation

You can include comments in your R package by using the roxygen2 package. First, you must install the package with the command install. packages("roxygen2"). This will install the roxygen2 library along with any supporting packages required.

The next step is to configure the build tools. Go to the build view and then choose "More" and then "Configure Build Tools...". You will see a dialog box like the one pictured in Figure 10-3 appear.

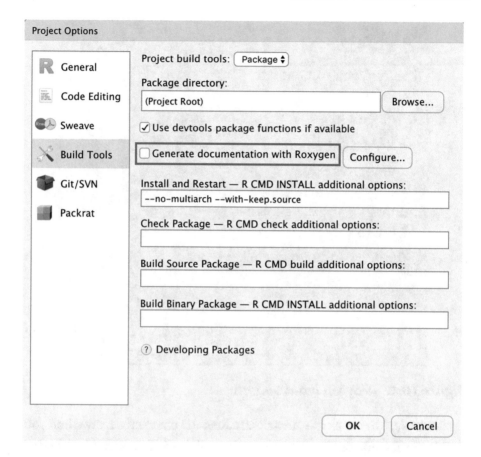

Figure 10-3. *Configuring build tools*

Click the "Generate documentation with Roxygen" checkbox. Then click the "Configure..." button to get the dialog box that appears in Figure 10-4.

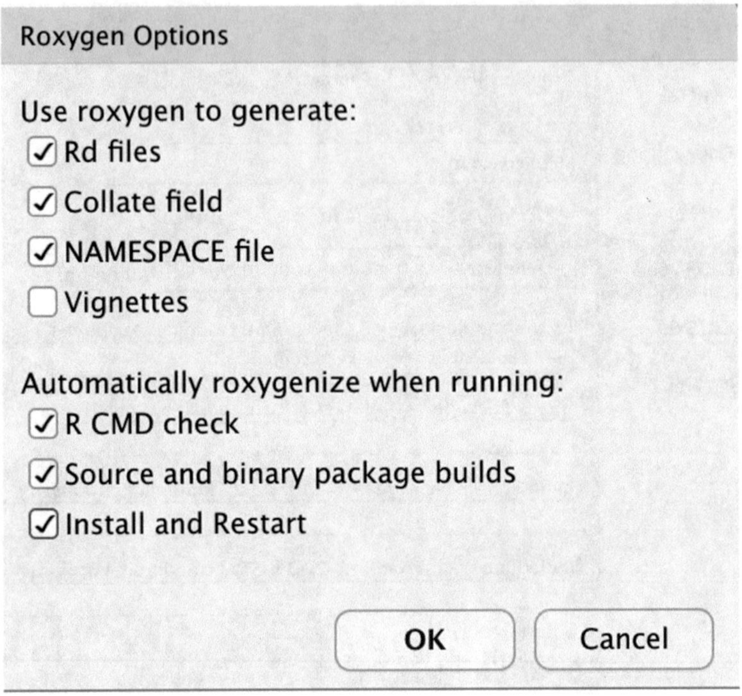

Roxygen Options

Use roxygen to generate:

☑ Rd files

☑ Collate field

☑ NAMESPACE file

☐ Vignettes

Automatically roxygenize when running:

☑ R CMD check

☑ Source and binary package builds

☑ Install and Restart

[OK]   [Cancel]

Figure 10-4. *Roxygen build options*

Make sure to check the "Install and Restart" checkbox. Now when you build your R package, the documentation will get automatically generated based on fields that you supply. You include documentation with special comments that include a backtick #'. Each field name begins with an @. Some common field types include @name, @title and @description. For example, here is what you might include in the hello function documentation:

```
# Hello, world!
#
# This is an example function named 'hello'
# which prints Hey there!.
#
#' @name hello
#' @aliases hello
#' @title Hello World
#' @usage hello()
#' @description  Prints 'Hey there!'
#' @examples hello()

hello <- function() {
  print("Hey there!")
}
```

Now when you build your R package your help documentation will reflect the content that you included above as pictured in Figure 10-5.

Figure 10-5. *Custom help documentation*

In the help file, we changed the description field to reflect that the hello world message was changed in the function.

Adding Datasets to R Packages

You can include datasets as well as functions in your packages. All you need to do is add a "data" folder to your R package and then put your datasets into that folder. These datasets can be R dataframes or raw datafiles of any type. For the life expectancy project, we simply copied the data folder that includes the csv file that was downloaded from the data.gov website. Then this code was used to save this dataset as an R dataframe.

```
library(readr)
life_expectancy <- read_csv("data/NCHS_-_Death_rates_and_life_
expectancy_at_birth.csv")

# Clean up dataframe
# Use shorter column names

names(life_expectancy)[1] <- "year"
names(life_expectancy)[2] <- "race"
names(life_expectancy)[3] <- "sex"
names(life_expectancy)[4] <- "life_expectancy"
names(life_expectancy)[5] <- "death_rate"

save(life_expectancy,
     file ="data/life_expectancy.rdata")
```

Next, click "Clean and Rebuild" in the Build view to rebuild the life package. Once this process is complete, you can access the dataset using this code from any RStudio project:

```
library(life)
```

```
head(life_expectancy)
```

This will show you the beginning of this dataset and more importantly, we can use this code from any analysis without re-running our life expectancy analysis.

Code Completion

As you type in code into RStudio, you may see that code suggestions will pop up with as you type. Code from your RStudio packages will follow this pattern as well. Also, you can quickly see what data and functions are available by typing in the name of the package followed by two semicolons as pictured in Figure 10-6.

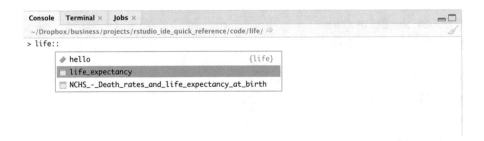

Figure 10-6.   *R Package code completion*

In the code above, we typed `life::` which is a way of filtering all the content that belongs to the life package. You can do this for any R package to see what functions are included with what packages. Your description field from your help documentation will also show up as you type as a tooltip.

Conclusion

In this chapter, we discussed how to create custom R packages to share our work with our team and beyond. R packages are a great way to encourage code reuse and provide a vehicle to document and package our work so that we can extend the life of our code beyond our own projects.

CHAPTER 11

Code Tools

RStudio includes tools that help you write, debug, and keep track of your code. We have already seen code tools that are built right into the RStudio editor such as code complete, syntax highlighting, and tooltips. We can also include git support to help keep track of versions of our code files, refactoring support to help edit our code, and debugging tools which help us step through and inspect code.

Source Control Integration

Source control is a system that you can use to keep track of changes that you make to your code files. RStudio supports integration with two source control providers: subversion and git. Both provide the essential feature of keeping track of changes you make through versions of your code.

If you already use source control, then you will appreciate the integration RStudio offers. However, if source control has never been a part of your workflow then the value of source control may not be immediately obvious. While it's a best practice to use source control, it is not a required feature of all projects. Still, it is a good idea to try to add this into your workflow.

© Matthew Campbell 2019
M. Campbell, *Learn RStudio IDE*, https://doi.org/10.1007/978-1-4842-4511-8_11

Setup Git

In this chapter, we will assume that you are using git and have this system installed on your computer. If not, please refer to Chapter 1 to install git. There are a few steps required to make sure RStudio knows how to use source control for you.

We need to indicate that we are using git in RStudio's preferences area. Go to the RStudio menu bar and choose "RStudio" and "Preferences" to get the dialog box that appears in Figure 11-1.

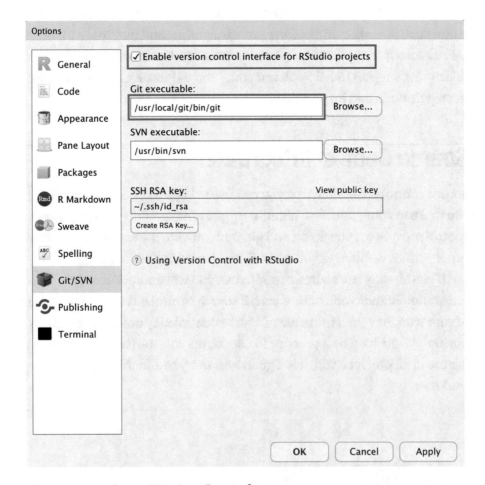

Figure 11-1. Setup Version Control

Check the box titled "Enable version control interface for RStudio projects" and make sure to provide the path to your installation of git. This path depends on your computer and where you choose to install git; the path shown is the default that folder for Mac installs.

Add Project to Git Repository

To add git to a project, you will need to use your Terminal by clicking the Terminal tab in RStudio. Then type in the command `git init` and press return. This creates a new git repository in the folder where our project is located. We can simply use git from the Terminal like this or with the integrated RStudio tools. While we are here let's add the files in this folder into our new git repository and then commit them:

```
git add .
git commit -m "initial commit"
```

These two commands add all our files into the repository and then commit them with the message "initial commit".

Git Viewer

Now, we will need to restart RStudio so we can see the git integration viewer appear as pictured in Figure 11-2.

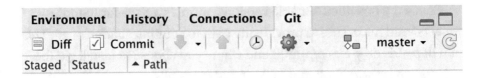

Figure 11-2. *Version control viewer*

When you make changes to a code file, the file will appear in this window. For instance, if I decided to make some changes to my "process_data.R" file you would see the source control viewer update to reflect this as seen in Figure 11-3.

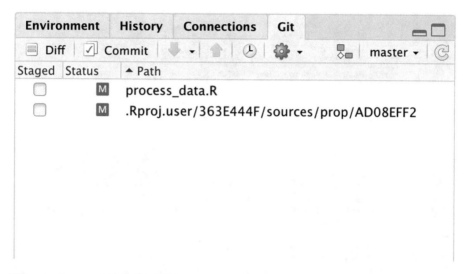

Figure 11-3. *Modified Files in Repository*

Now we can see that two files appear in the git viewer. We expected "process_data.R" to appear, but not the second file with the long file path name. The blue "M" icon next two these two files means that these files were modified. Let's take a look at the changes in the first file.

Git Diff Viewer

Click the checkbox next to each of these files and then click the "Diff" button in the top left area of the viewer. A new screen like the one pictured in Figure 11-4 will appear.

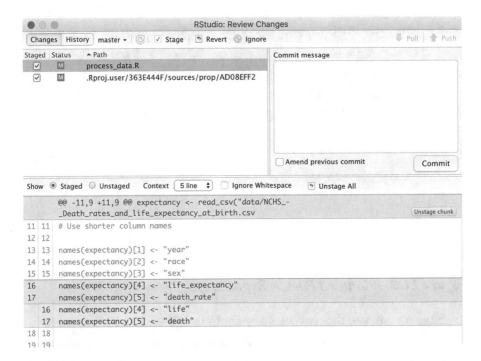

Figure 11-4. *Diff Screen*

On this screen, you will be presented with the differences in the files. The white space shows the content that are in both files. The red space shows the content in the previous version of the file (that is about to be deleted) and the green content shows the new additions to the file.

Use the list in the upper left-hand area to navigate between the two files that are changed. If you looked at the other file, you would see that the changes have to do with the state of our code editor which is not that important and can be ignored (we will see how to ignore files in a moment).

If you click the "History" button at the top, you will see the history of all changes made to this file since we started working on it. From this point in the workflow, we can either get rid of these changes or we can commit the changes to our repository.

Committing Changes

To commit the changes, we can type in a "commit message" to in the upper right-hand area of the screen and then click the "Commit" button as pictured in Figure 11-5.

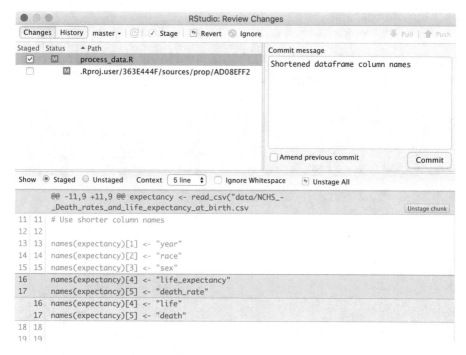

Figure 11-5. *Commit changes to the repository*

We could have removed the changes by clicking the "Revert" button instead of the "Commit" button.

Git Branches

Git has support for branches. You can create a branch when you want to make changes to a copy of your project so that you can test new features without affecting your original code. If you work on a branch and then later decide that you would like to keep the new feature, you have the ability to "merge" the branch into your original project.

You can name new branches anything that you want, often you might use the name of the feature or bug you are working on. Your original source branch is named "master". The git workflow that you generally follow is to create a new branch from the master branch, work on the new branch and if the new branch meets your requirements you will merge it back into the master.

RStudio includes support for creating new branches and switching between branches. Use the buttons highlighted in Figure 11-6 to add new branches and to switch branches.

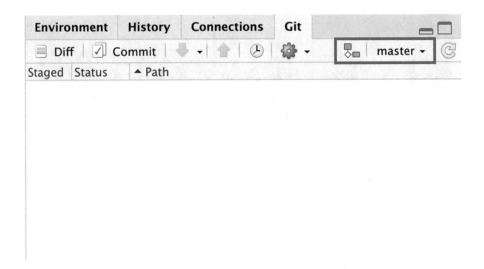

Figure 11-6. *Managing git branches*

The first button is used to create branches and the second is used to switch branches. When you switch branches all the code in your project will be replaced with the code from the branch.

You can get a picture of all the branches in your repository by clicking the "Diff" button, then the "History" button, and then choosing "(all branches)" in the space right next to the "History" button. In Figure 11-7, an example of a project with a master branch and a "new_feature" branch is shown.

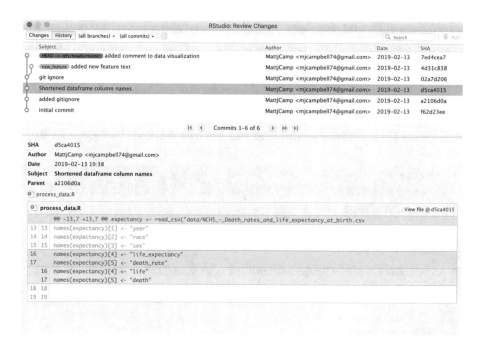

Figure 11-7. *Git branches*

You can use this tool to navigate through and visualize the differences in your branches and to see the history of changes in the project. RStudio does not have support for all git features. If you want to delete branches or merge branches you will need to use the Terminal tab and look up the appropriate commands.

Debugger

Debugging is the process of identifying and fixing defects in your code. The simplest way to debug may be to simply print out object values to the Console screen. However, this can become frustrating when the state of your analysis changes as each line of code executes.

To help with that problem, RStudio comes with a built-in tool that lets you set breakpoints which will stop your code from executing and give you a chance to inspect object values.

Use the debugger by setting breakpoints. Simply open a code file and click in the area to the left of the line of code that you would like R to stop at. This area is called the gutter and you see this pictured in Figure 11-8.

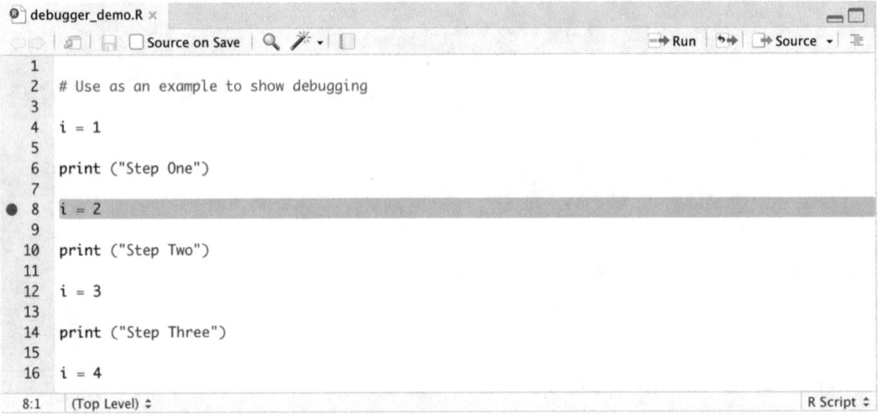

```
  debugger_demo.R ×
         Source on Save   Q                                    Run      Source
  1
  2   # Use as an example to show debugging
  3
  4   i = 1
  5
  6   print ("Step One")
  7
● 8   i = 2
  9
  10  print ("Step Two")
  11
  12  i = 3
  13
  14  print ("Step Three")
  15
  16  i = 4
  8:1    (Top Level)                                              R Script
```

Figure 11-8. *Setting a breakpoint*

The red dot next to the i = 2 line of code is the breakpoint. When we execute this code, RStudio will stop at this line of code and present the code debugger as pictured in Figure 11-9.

Figure 11-9. *Key debugger features*

In Figure 11-9 you can see all the debugger features. In the middle-left area, we have a set of controls that we can use to control the code execution. Now that we are stopped at the breakpoint, we can advance to the next line, continue all code until the next breakpoint, or stop code execution.

Toward the bottom of the screen, we can use the Console to write code to inspect the current state of the project. This is a nice feature when you are working on testing ideas about code defects. You also have the environment screen at the top where you can see the values of all the objects in the environment.

You get a few more options for debugging by going to the RStudio debug menu. The first two options in particular are helpful and let you toggle and clear your breakpoints.

More Code Tools

RStudio also comes with a few more tools to make your code editing and refactoring a little bit easier. Code refactoring is something you do once you have a working project, but you want to clean up your code a bit. You may want to change function names, indent your code or extract regions of code into re-suable functions.

Go to the RStudio header and choose the "Code" menu item to see all the options you have to help refactor your code. One key item is the ability to "rename in scope", which you use by highlighting an object and then selecting the menu. At this point any edit you make to the object will be reflected in all appearances of this object in the code file.

You may also find the option to re-indent lines and reformat code helpful to give your code an organized look. Navigating code is much easier when you use the "Go To Function Definition" when you have a called function but you can't quite remember where it was defined.

Conclusion

In this chapter, we highlighted some built-in tools that RStudio comes with that will help you organize and maintain your code. Git is used to organize and see differences in versions of your code. You use the debugger to observe the state of your program as it steps through code so you can fix bugs. Finally, the code menu gives you a list of utilities that you can use to reformat your code quickly.

CHAPTER 12

R Programming

In this last chapter, we are going to go over the basic concepts that you need to know to do R programming. This tutorial is in no way comprehensive or meant to be your first experience with programming. However, if you have programmed before, this should provide you with enough to get started in R. Readers that are completely new to programming may want to consider taking an intensive course on this subject perhaps with Data Camp or Coursera.

Objects

Everything in R is an object. An object can be a list of numbers, a function, a plot, or the result of an analysis. There are different types of objects such as lists, data frames, and characters.

You create an object with the assignment operator <-. On the left is a name that you choose, followed by the assignment operator <-, and then code that defines the object. You create a numeric object like this:

```
a <- 9
```

The first part a is the name we choose for the object, followed by the assignment operator <-, and then a value 9. We will refer to this object in code now by just using the letter a. For instance, if we wanted to know how to double the value of a we could do this:

```
a * 2
```

© Matthew Campbell 2019
M. Campbell, *Learn RStudio IDE*, https://doi.org/10.1007/978-1-4842-4511-8_12

This would print 18 to the Console. Objects can be different types in R. The object a above is numeric. These types are called classes in R. You can find out what the class of an object is by using the class function.

```
class(a)
```

This will return the value numeric to your Console. You can change the class of an object with the assignment operator like this:

```
class(a) <- "custom_numeric"
```

Generally, you will not change the class of objects that already exist like numeric or character. However, you may if you are creating your own custom class and want to distinguish your objects from others present in the environment.

Essential Class Types

We can define and use multiple types of objects while writing code, but there are a few which are foundational to R programming, so these will be covered specifically below.

Character and Numeric

We covered numeric objects in the first section and character objects work in the same way. The real difference is that character objects store characters and not numbers. Character objects need to be wrapped in single or double quotes like this:

```
b <- "nine"
```

When we reference b in code, we will get the string "nine" returned. Unlike the numeric object we looked at before, we cannot multiple b by 2. If we attempted this, we would get the error "non-numeric argument to binary operator".

While we cannot do math on character objects, we can use the `stringr` library to manipulate these objects. For instance, if we wanted to combine two character objects together we could use the `str_glue` function from the `stringr` library like this:

```
library(stringr)
str_glue(b, " times 2")
```

This would return the string "nine times 2". The `library` keyword is used to give us access to a library of functions. The first line of code above means that we need access to the `stringr` library.

Vector

A vector is a list of objects of the same type. You may have a list of character, numeric, or other type of objects. You keep track of objects in the list using a numeric value starting with 1. To create a vector, you use function named c that takes a comma-separated list of values to create the list. Here is how you create a vector containing the letters that make up the first five letters of the alphabet:

```
alpha <- c("A", "B", "C", "D", "E")
```

You could simply print the entire list out to the Console by typing in the name `alpha` or you could reference individual elements by using the object name followed up square brackets with an index that corresponds to the element in the vector that you are interested in.

For instance, if you `alpha[1]` will print "A" while `alpha[5]` will print "E".

When you apply an operation like multiplication or a function to a vector the operation will get applied to each object in the vector. The result will be a new vector. For instance, if we had a vector of the numbers 20, 30, 40, 50, and 60 and then multiplied the vector by 4 we would end up with a new vector 80, 120, 160, 200, and 240.

```
n <- c(20,30,40,50,60)
n
[1] 20 30 40 50 60
n * 4
[1]   80 120 160 200 240
```

Matrix

A matrix is a vector that has two dimensions (essentially rows and columns). Like a vector, each object in a matrix must be the same type. You can create a matrix by setting the dim (for dimension) property of a vector like this:

```
m <- c(1,2,3,4,5,6,7,8,9)
dim(m) <- c(3,3)
m
```

When you inspect m you will see the matrix:

```
     [,1] [,2] [,3]
[1,]   1    4    7
[2,]   2    5    8
[3,]   3    6    9
```

To reference individual cells in the matrix, you supply the row and column index. For instance, to see what is in row 2 and column you would do this to m[2,1] to get the value 2.

List

The vectors we discussed before are called atomic vectors because they must contain all the same elements. Vectors that contain mixed objects are simply called lists. Lists are handy when you want to manage collections of related objects.

To create a new list you can just the list function like this:

```
l <- list()
l$name <- "Cards"
l$cards <- c("10","Jack","Queen","King","Ace")
l$bets <- c(1,5,10)
```

The first line creates an empty list and then uses the $ to create names to assign objects to. The first is a character object that describes the list, the second is a vector of characters, and the last is a vector of numeric objects. To access these values later on you must use the list name followed by the dollar sign $ and the name of the object. For instance, to see the list of cards you would type l$cards and to get the second card in the list you would type l$cards[2].

Factor

Factors are type of vector that can only include specified values. For instance, you may have a list of values to indicate sex and each value must be either "male," "female," or "unknown." This structure is handy in those situations when we are doing analysis on categories of objects.

You create a factor using the factor function and supplying the list of values.

```
sex <- factor(c("male","female","unknown","male"))
sex
```

This code outputs this information:

```
[1] male    female  unknown male
Levels: female male unknown
```

We have a list of values like we would in a vector, but we also have additional levels information.

Data Frame

A data frame is a special kind of list that must have components of equal length. In effect, this provides a sort of spreadsheet like two-dimensional data structure comprised of rows and columns. You can treat data frame components like a list with the $ operator or like a matrix using row and column indexes.

You can create a data frame using the data.frame function:

```
df <- data.frame(numbers = c(1,2,3),
                 letters = c("A","B","C"))
df$more <- c("1st","2nd","3rd")
```

You can supply vectors all of equal length to the data frame when you create it with the data.frame function. Once you have a data frame you can create additional columns using the $ operator as you can see above.

Oftentimes, you will have a data frame created for you as output from a function. Data frames are an essential R object and the "tibble" version type of data frames are covered in great detail in the "essential R packages" chapter.

Flow Control

Up to this point, we have essentially created objects, set properties, and inspected the results. R programming also includes methods to branch code and to repeat code using loops.

If-Then Statements

An if-then statement instructs our program to take different actions depending on the state of certain objects. Here is an example of an if-then statement:

```
state <- "startup"

if(state == "startup") {
  print("We are about to start the program")
  state = "running program"
} else {
  print("Program running")
}
```

We have an object named state with the value of "startup". The next line of code evaluates the state object and if the object is equal to "startup" then the code right after the if statement in the curly brackets will execute. A status message will print and the value of the state object will change.

If the state object is anything other than "startup" the code after the else keyword will execute. Here the program would simply print out the message "Program running".

In the example above, we used the == operator to test whether the two values are equal. You can also test for inequality ! =, greater than >, less than <, greater than or equal to >= and less than or equal to <=.

You can also use if-else on a single line of code and this works well if you have concise code. For instance, if we just wanted to test the state object to see if it was in startup mode we could simply do this:

```
ifelse(state == "startup", TRUE, FALSE)
```

If state is "startup" then we will get the Boolean value TRUE. Otherwise we get the Boolean value FALSE. Booleans are a special type of object that can only be true or false and are used to test conditions like this.

Loops

Loops give us a way to repeat the same code over again. There are two types of loops that we can use in R: the for loop and the while loop. We also use the apply family of functions that provide loop-like behavior. Let's see how loops work.

For Loop

For loops will execute a specified number of times. For instance, if you want to loop through code five times you can do this:

```
for (i in 1:5){
  print(i)
}
```

This will print out the numbers 1 through 5. This code is made up of the for keyword and in parenthesis we have an object i that keeps track of where we are in the loop. 1:5 is R shorthand for every number 1 through 5. This loop will run 5 times and the code between the curly brackets will run each time producing this output:

```
[1] 1
[1] 2
[1] 3
[1] 4
[1] 5
```

While Loop

While loops also execute code repeatedly. But, instead of executing a fixed number of times these loops run until a condition is met. This is how we would count from 1 to 5 with a while loop:

```
i <- 0

while (i <= 5) {
  print(i)
  i <- i + 1
}
```

The while loop checks the state of the i object and it prints out the value to the Console until the condition i <= 5 is false. Each time the loop iterates the value of i increases by 1.

Apply Functions

The apply family of functions will also provide the looping pattern but in a more compact form. You will find this type of function used in various R packages and also in base R. The general pattern is that you use the apply function by passing an object such as a vector, data frame, or list and the apply function will iterate through each element in that object. You provide the apply function with code to execute against each element and the function returns a new object.

To replicate what we have been doing so far with loops, we could do this:

```
sapply(1:5, function(x){print(x)})
```

This is much cleaner, and we only really have the vector 1:5 and the code to execute. Note that the code is an "anonymous function" and functions will be discussed in more detail next. Basically, we have a function defined that has one parameter x which is a placeholder for the current value in the vector, and in between the curly brackets we have the code that will execute for each element in the vector.

We used the sapply function here and this version of the apply function returns a simplified object (in this case a vector). In the case above, we just get the same vector back. However, we could have done something more like this:

```
sapply(1:5, function(x){
  x * 2
})
```

We broke out the anonymous function here to make the code a little bit more clear. This code will produce a new vector that will print this to the Console:

```
[1]  2  4  6  8 10
```

This apply function is a general wrapper function that returns the simplest object. We also have `lapply` which is used for lists and data frames that work the same way but return list objects.

Functions

Functions are a type of object that contain a block of code that can be executed with inputs that you provide to the function. These objects are a great way to follow the best practice of writing code once. Generally, any block of code that you feel can be used throughout your project can be wrapped up as a function to be used later on. Here is how you create a function object:

```
my_fun <- function(x, y) {
  r = x * 2 + y
  r
}
```

The code in between the curly brackets { and } is what makes up the function. Inside these curly brackets, you can create new objects and reference objects that are declared as parameters in the function. Parameters are declared in the comma-separated list that appears in parenthesis after the function keyword. Functions can also use objects that have been declared outside of the function as long as they have been declared before the function itself.

The function including the parameters list and code block is assigned to the my_fun object. We can inspect the function by typing the object name into the Console to get a printout of the code in the function.

In order to use the function, we must type out the function name followed by the comma-separated list of parameters like this:

```
my_fun(3, 189)
```

This will return the numeric object 195. What happens in the function is that each line of code is executed, and the last line of code is returned to the caller. The last line of code in my_fun is the result of the calculation.

Functions that are not assigned to object names are called anonymous functions and are often used as parameters to functions like lapply or sapply. These functions work in the same way but are not necessarily reused in the same way as a standard function would be.

Importing JSON Data

JSON, JavaScript Object Notation, is a data format that some programmers may not be as familiar with as compared to csv or other two-dimensional data formats. The JSON data format is designed to be something that can be sent easily in a text format while maintaining a structured format.

Here is an example of JSON data:

```
[{"year":2015,"race":"All Races","sex":"Both Sexes","death":
733.1},{"year":2014,"race":"All Races","sex":"Both Sexes",
"life":78.9,"death":724.6},{"year":2013,"race":"All Races",
"sex":"Both Sexes","life":78.8,"death":731.9},{"year":2012,
"race":"All Races","sex":"Both Sexes","life":78.8,"death":732.8},
{"year":2011,"race":"All Races","sex":"Both Sexes","life":78.7,
"death":741.3},{"year":2010,"race":"All Races","sex":"Both
Sexes","life":78.7,"death":747}]
```

This is a portion of the life expectancy data that we worked with throughout the book. You can probably make some sense of this data, but it's far from a friendly format.

To work with data like this you can use the jsonlite package. Install this package with this code:

```
install.packages("jsonlite")
```

Now, assuming we have the JSON data above stored in an object called named life_json we can convert the data into a tidy format that we can use with this simple line of code:

```
library(tidyverse)
library(jsonlite)
life <- fromJSON(life_json)
```

We included the library references above. You can inspect the result by typing the object name life into the command line to get this data frame:

```
year        race        sex death life
1 2015 All Races Both Sexes 733.1    NA
2 2014 All Races Both Sexes 724.6 78.9
3 2013 All Races Both Sexes 731.9 78.8
4 2012 All Races Both Sexes 732.8 78.8
5 2011 All Races Both Sexes 741.3 78.7
6 2010 All Races Both Sexes 747.0 78.7
```

The jsonlite package fromJSON function automatically transformed the file into a data frame. This function will return the most high-level object that it can automatically. Since our data fit neatly into a data frame that is what we received. However, if the underlying data frame does not fit into this structure you may end up with a list object that reflects the structure of the JSON file.

Conclusion

In this chapter, we have presented enough information about R programming to get you started and to make sure that you can follow along with this book. You should now understand objects, control flow, and functions in R. Additionally, the chapter on essential R packages and the section on JSON will give you a solid start if you have produced data analysis in the past.

For readers new to data analysis and/or programming, you may find that you need more help. Programming is something that takes a lot of practice and some understanding in the beginning. You may benefit by reviewing other programmers' data analysis work. Many examples are available online on websites like Github.

Readers who are completely new to programming may want to take some time to learn essential programming. If this book is the first time that you have encountered ideas such as loops, functions, and objects then you should consider taking a local or online course on R programming.

For everyone, we hope that the material in this book and this chapter will help you understand what tools are available to you so that you can really start to provide the insight into your work that you are looking for. Good luck and I hope that you enjoy the journey into the technology of data science!

Index

Printed in the United States
By Bookmasters